T0251900

Programming GPS and OpenStreetMap Applications with Java

Programming GPS and OpenStreetMap Applications with Java

The RealObject Application Framework

Kristof Beiglböck

CRC Press
Taylor & Francis Group
Boca Raton London New York

CRC Press is an imprint of the
Taylor & Francis Group, an **informa** business

AN A K PETERS BOOK

CRC Press
Taylor & Francis Group
6000 Broken Sound Parkway NW, Suite 300
Boca Raton, FL 33487-2742

First issued in hardback 2017

© 2011 by Taylor & Francis Group, LLC
CRC Press is an imprint of Taylor & Francis Group, an Informa business

No claim to original U.S. Government works

ISBN 13: 978-1-138-41374-0 (hbk)
ISBN 13: 978-1-4665-0718-0 (pbk)

Library of Congress Cataloging-in-Publication Data

Beiglboeck, Kristof.
   Programming gps and openstreetmap applications with Java : the realobject application framework / Kristof Beiglboeck.
      p. cm.
   Includes bibliographical references and index.
   ISBN 978-1-4665-0718-0 (alk. paper)
      1. Global Positioning System--Computer programs. 2. OpenStreetMap (Project) 3. Java (Computer program language) I. Title.

   G109.5.B45 2011
   910.285'5133--dc23   2011043869

Visit the Taylor & Francis Web site at
http://www.taylorandfrancis.com

Contents

Preface

Software development has come a long way from data structures to objects. The entity of an object has changed the perception and design of programs. Objects have become independent programs separating external meaning and internal processing. According to the natural meaning of "object" derived from physical things, software objects were initially introduced to reflect real-world representations. This book traces the idea of creating *real objects* as virtual projections of physical things.

Virtual machines were introduced to abstract the underlying hardware and to free object development from operating systems. A modern GPS smart phone represents a personal programming environment equipped with location and other sensors and broadband connectivity that can submit and retrieve real-world information in real time. The book suggests a Java implementation for remote objects on mobile runtime environments. Each object can collect telemetry of its real-world counterpart to become an element of a large network. For example, imagine a rudimentary smart phone attached to a car, collecting engine characteristics together with its GPS trace and submitting it live to a real-object server application.

The increased availability of live data poses a great challenge to reflect the real world in a virtual world. Digital maps have gained importance and are an essential tool for organizing spatial data. The processing of digital maps is described in detail in this book. Today's GPS devices represent a fascinating and relatively cheap technology for scanning the planet. To connect server and client objects in a semantic way, the book covers the handling of GPS data as client coordinates for time and space.

We develop a general multipurpose architecture to deal with GPS data from many client objects at a time. The intention is to unify and reduce the programming effort with a Java framework to coordinate (live) GPS data on digital maps. A *real-object application* is created as a server software to collect live data from any number of remote real objects. The server software is pre-implemented to synchronize and validate the remote objects' behavior. As the server is designed *not* to distinguish real from simulated data, simulated clients can enrich the real-world scenario as long as they behave realistically and don't raise plausibility conflicts. A game is developed as an example application; the client objects (in this case, the players) will be enabled to navigate along the network of a server map, while making intelligent decisions in the context of the game.

Generally speaking, each server represents a real-world scenario for real objects to act and interact in. The reader can configure a server for his dedicated real objects and extend the implementation with domain knowledge. Every real-object application represents a controller for a specialized class of real objects. Vice versa, the server validates the remote data of its authorized clients, approves plausible behavior, and can provide dedicated client data to the *real-object application framework*. The fractal architecture of the real-object application framework is designed to combine any number of real objects and real-object applications (of many users) to reflect as much reality as possible. By nature, every aspect of reality can be validated against and combined with other realistic scenarios without contradiction—to form a real-world simulation.

Technically speaking, this book describes the process of "modeling the world" by implementing a minimum reference real-object application to initiate life-cycle development in a community process. The book demonstrates how to set up a first life cycle by using unified modeling to concrete implementations in Java. While this book does *not* teach programming from scratch, it illustrates the essentials of object-oriented concepts using a single Java project as an example. The project development is aligned with the online Java Tutorial and can serve as a tutorial for preparing for the Java Certification exams.

THE JAVA TUTORIAL > TUTORIAL TRAIL > SECTION

 Gray boxes point to Java skills which may be helpful as you proceed through the text and your understanding of Java. The Java Tutorial can be found at `download.oracle.com/javase/tutorial`.

While the Java Tutorial examines a collection of code snippets to illustrate the semantics of Java, this book shows how to build *one* application to convey the essential idea of modeling a task through the Java objects. The source code and sample application are available to the reader on the book's website `www.roaf.de`. The book should be used as a hands-on instruction accompanying the source code. The reader is expected to download and compile the sources in a personal development environment (i.e., eclipse).

Double-framed boxes indicate the hands-on instructions to be executed while reading each development step of the distributed application. In this way the reader can immediately experience how code changes influence the entire system. As the skill set is increased in Part IV, it is necessary to study and run the code in order to follow the server development.

Please read the download instructions in the Appendix.

This book also provides readers the opportunity to learn in groups. An instructor can begin a workshop by having students play against each other remotely on a game server. The group can be split into different functional areas (i.e., GUI, client, server) and experience distributed-application development. Each student can develop his own real (world) object according to his skill set and participate in a scenario defined by rules. Implementations can range from very simple code all the way to artificial intelligence, depending on individual knowledge, skills, and interests. Testing and verification implicitly takes place by interacting online with other (students') real objects.

I invite you to become part of the evolution from Web 2.0 to a real-time Web 3.D !

Part I

From Vision to Mission

This first part of this book takes a thorough look at the early intentions of object-oriented programming and the implications thereof. Java can be seen as the first programming language consistently developed with software objects. This evolutionary step, wrapping up three decades of experience and experiments with software objects becomes apparent by comparing objects to independent electronic devices.

It is barely possible to find an introduction to software objects without a reference to the real world. This book thus uses the analogy of objects representing real physical objects in general and in detail. A look at models in physics will help to grasp the terms "real" and "physical" to introduce the idea of *real-world modeling*. Physical laws are represented by mathematical formulas for dedicated coordinate systems and real-world values.

From a good understanding of software objects, devices, and physical things, a vision is derived to create a virtual world reflecting the real world. A modern GPS smart phone makes it easy for anyone to implement a client software object to collect real-world data and transmit it to a server. The server application can validate different client data against each other and against third-party sources to serve as a reliable source of bundled *real-world data*.

Naturally, the vision of a *real-world simulator* leaves plenty of room for discussions and raises different expectations. Therefore a rough object-oriented analysis (OOA) is used to formalize the vague vision into a software architecture to enable distributed development of the individual components. The mission of this book is to discuss the idea of a REAL OBJECT APPLICATION FRAMEWORK and provide a "minimum implementation with maximum abstraction." This implementation should encourage the reader to develop his own scenario and establish a life cycle development of the `roaf` library in a community process at the website `www.roaf.de`.

Chapter 1

Software Objects and Real-World Representations

1.1 Introduction

This chapter draws an analogy between objects in software development and electronic devices, and to physical objects in general.

remote control Objects can be controlled by an application via application programming interfaces (API) similar to controlling a television set via a remote control. Every button invokes a "method" on the electronic device, while the functionality is hidden.

We introduce a video application to demonstrate the difference between developing programs and developing objects. The virtual machine represents the actual programming environment referencing (external) objects with built-in functionality. This demonstrates one design intention of the book, to make object and application development as independent as possible.

This method is not limited to devices—any other technical construction can be modeled by a software object. For example, a car is an object that can be perceived externally, operated internally from the driver's seat, and tuned internally under the hood. Programming a car should be closely related to building a car in order to reflect real car behavior. Then, a traffic application can make use of a car class to instantiate a number of cars.

In the development of our methodology, we look at how scientists model the world. Physicists have actually re-engineered the behavior of physical objects in general—not only man-made devices and machines. The external characteristics of a stone are simple to describe. Yet, it took centuries to understand its inner structure and its behavior, for example, in a gravitational field—the external environment. This behavior can be simulated with mathematical formulas.

Figure 1.1. The programming environment can be compared to a programmable universal remote control.

1.2   Controlling Remote Objects via References

Object orientation has been an evolving principle in software development for more than three decades. Java, developed by James Gosling and released in 1995, derives much of its it syntax from C and C++, but has a simpler object model. Java was originally developed as a language to control *devices*—not primarily computers. This initial intention can help us to understand the entity of "objects" by drawing the analogy between software objects and hardware devices.

A television set is a sophisticated technical device and its "users" don't care much about *internal* functionalities. Its primary functions are *abstracted* with the buttons of the remote control. The haptic remote control (HMI) serves as the human-machine *interface*, similar to a graphical user interface (GUI) on a computer screen. The buttons of the remote define a set of commands (API) and variables to *control* the device:

- one button to wake up the television and turn it on and off
  `boolean on/off`;

- one element to switch channels (controlling the receiver *component*)
  `integer 1-50`;

- one element to control the sound (amplifier and speakers)
  `integer 0-100 (%)`;

- two elements to control the brightness and color (display)
  `integer 0-100 (%)`.

From a programmer's point of view, the remote control is the programming environment and the TV set is the controlled *object*. The program has a *reference* to the TV realized by wireless transmission. Many televisions are controllable only via remote control and without it you loose the *reference* to the *object* television—the TV would go into the "garbage collection"! An object is useless, if it can't be referenced.

Object referencing also raises awareness of a very important aspect: Imagine you and your neighbor have identical TV sets. If you have two programs (controls) referencing two devices, how do you prevent your neighbor from changing the *state* of your TV, for example switching your channels.

1.2.1 A Video Application

In order to set up the programming example, the television will be divided into *components*, and three multimedia devices will be added:

- The TV set is composed of

    - a screen (the front end),

    - a tuner to receive external signals from an antenna, cable, satellite dish, etc.,

    - a sound device (amplifier and speakers);

- a hard disc to record the signal from the receiver (persistence);

- a DVD burner to record the signal of a selected channel or from the hard disc and a second tuner to receive additional signals;

- a clock (signal).

(One might consider the clock as a background synchronizer for the actual devices and in a way, part of the program environment.)

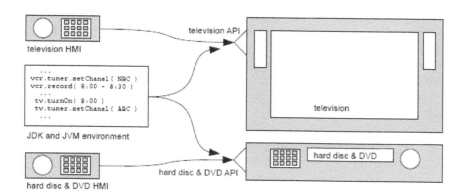

Figure 1.2. The java programming environment can control any device as long as it supplies a java application interface (API). One application can replace all remote controls, if it can access or reference the individual devices APIs.

In this context, we can describe the initial intention of Java—to provide the programming environment for a multimedia application and abstract each component to its core functionality (see Figure 1.2). A programmer of a multimedia center would then be able to replace any component with another one, and hook it up to the application via a standard interface— without modifying the controlling application. He can program a higher-level *video application*. We consider this application as higher level, since it combines steps that would normally require input on every single remote control of the various components.

The entire application can be seen as a highly flexible remote control and can be implemented on some hand-held device (with the ability to host a Java Virtual Machine (JVM)). Java makes it possible to code *the logic* of any application in a neutral (or standardized) programming environment.

In this section, we walk through a video application using object-oriented terminology with *devices* being special types of objects or components. The purpose of the video application is to provide a consumer friendly front end with high usability. This example gives an introduction to programming with objects.

For the initial analysis, the following sample *use case* will be used: the user would like to record tonight's ball game, without knowing the kickoff time, and watch the news at eight.

Application programming interface. By defining the API for the functionality (or *services*) of each component, the complete functionality is defined. At this point the individual APIs are not listed; they will become apparent in the program flow.

Application program flow. The program flow is the environment where objects are *referenced*, controlled, and combined in a logical context. In order to execute commands (*methods*) on a device, it has to be powered up and initialized. In the context of object-oriented programming, we *construct* objects with the **new** operator and the *constructor* of a class. In a way, the power button represents the constructor of a device!

The beginning of an object-oriented video application could look like this:

```
tvGuide = new EPG( myZipCode )
     tv = new TelevisionSet( modelXY )
    vcr = new VideoRecorder( modelYZ )
```

These three lines of code have many implications to *distributed* operations: Each line might wake up a device, do a system check, or start up the subcomponents. These operations can take time, and the constructor of each component should be designed in a way that it returns some error code, if something goes wrong, e.g., if an *exception* occurs. Nevertheless, the order of *creating the objects* does not imply that they will be ready at the same time. After this initial phase the *hardware components* are powered up and in a state to process commands, that is, the *software objects* are *instantiated*.

The program (environment) *using* the new objects could continue like this:

```
Show ballgame = tvGuide.find( "NFL basketball", TONIGHT )
Show     news = tvGuide.find( "news", "8:00pm", "CBS" )
```

The program creates two `Show` objects describing a "show" with attributes: `starttime`, `endtime`, `channel`, `name`, `description`, ... The electronic program guide (EPG) implies an online connection to a current program guide and can be used to look up and return certain shows by using the *method* `find`. The EPG API offers the method with different *signatures*:

- `find( showname, showtime )`
  The EPG will be searched for the entry "NFL basketball" in the predefined time range `TONIGHT` on `TODAY`s date.

- `find( showname, showtime, channel )`
  This find narrows the search by providing the specific channel to search.

By providing these `find` methods as part of an API, the actual searching of an entry is completely hidden from the programmer of the video application. After having found the `ballgame` and the `news` items (with duration and `stop time`), the application can utilize them in the next steps of the programming logic:

```
wait until (ballgame.starttime == clock.now )
then do
    vcr.Tuner.setChannel( ballgame.getChannel )
    vcr.record( vcr.Receiver.getSignal )
        until ballgame.stoptime
```

The above code snippet implies yet another object and device: a clock. The clock component is used by a timer to compare the `starttime` of the `ballgame` with the current time. When the ball game starts, a receiver is tuned to the ball game's channel and the stream is directed to be recorded on hard disc until the ball game's end time is reached.

A second loop is waiting for the news at eight and will turn on the television and set the news channel. The waiting loops should run independently of each other, and the television should have its own receiver, so as not to disturb the recording of the ball game.

```
wait until (news.starttime == clock.now)
then do
    tv.turnOn
    tv.Tuner.setChannel( news.getChannel )
        until news.stoptime
```

The pseudo listing shows how distributed components can be represented by objects. Object references in the program flow are like remote controls for devices. The program has abstracted technical details away from the semantic programming and separates objects from program flow![1]

THE JAVA TUTORIAL > LEARNING THE JAVA LANGUAGE > WHAT IS AN OBJECT?

...Look around right now and you'll find many examples of real-world objects: your dog, your desk, your television set, your bicycle. ...They all have state and behavior. ...Identifying the state and behavior for real-world objects is a great way to begin thinking in terms of object-oriented programming.

...your desktop radio might have additional states (on, off, current volume, current station) and behavior (turn on/off, increase/decrease volume, seek, scan, tune).

Our video application demonstrates the mechanics of object-oriented applications, where, in our case, the objects are electronic devices. The objects being instantiated are generally not part of the object-oriented language being used. Actually every device should be distributed with associated classes, similar to a hardware driver. The vendor of a hard-disk recorder could develop the object `HDRecorder`, while the vendor of the video application uses an object as if it were an HD recorder. Objects can be seen from their internal functionality or externally, interacting with other objects in the context of the application.

The JVM is the vital component of every Java distribution: it abstracts the actual hardware and provides a programming environment to execute the programming logic on referenced objects. Java offers the runtime environment to create, observe, and control any number of objects/devices at the same time and transmit messages/signals between them—even if they reside on different machines.

The JVM, however, only handles language keywords, numbers, Boolean values, characters, and objects. This implementation of an architecture-neutral and portable language platform makes it easy to improve code, without getting involved with higher-level language constructs.

---

[1]Note: The gray boxes point to Java skills which may be helpful as you proceed through the text and your understanding of Java. THE JAVA TUTORIAL can be found at `download.oracle.com/javase/tutorial`.

1.3 Object-Oriented Programming

The video application served as a simple example to get a feel for object-oriented programming (OOP) by referencing (or projecting) a real-world object (or device) into a software environment. Getting used to thinking in OOP terms, basically means recognizing scenarios as a communication (or exchange) between objects, or things. It is important to understand, that object-oriented programs mainly handle and use objects. The objects themselves are usually designed separately (like devices) and often with different intentions and for various environments.

After we used electronic devices to introduce the idea of independent objects in a common programming environment, we now turn to a class of objects that is a primary focus of this book: a `Car`. What is a car? It is not surprising that asking different people this simple question in different situations results in different answers.

A `Car` can be a part of many scenarios:

- A passenger of an airplane perceives a car as a tiny rectangle moving on a line (road).

- The passenger of a cab only has to know how to call one, get in, and give the destination to the driver.

- The driver of a car has access to more private "methods" to control the car's behavior:
  `startEngine - shift - throttle - break - steer`

- The mechanic uses more specific methods to open the hood to fix or tune the machine, which is usually done in the garage (develop) before the car is being driven (deploy).

Nevertheless, one car is always the instance of one object (model), no matter how you look at it or where you drive it. In OOP terminology, each scenario (application) can require a different abstraction level. One idea of programming with objects is to create objects with high abstraction (e.g., rectangle moving on a line) and actually use these simplified objects in selected applications.

The object `Car` is an isolated entity. And just like real cars have been improved in every aspect over decades, the object `Car` can be enriched for every new application in which it is used. This style of programming automatically leads to *code reuse* for all aspects of a car in one place. This process of learning, testing, and reusing isolated objects makes OOP so robust, even in unexpected situations.

In what follows, we develop objects and applications as independent of each other as possible. For example a very simple class `Car` can be created to instantiate a number of `Cars` for a simple application. To show a rectangle (representing a car) moving on a map to display a `Car` can be implemented with only a few details or attributes like `size`, `location`, `speed` and `direction`. A `Car` implementation with an internally hidden wireless connection to a receiver could pick up a real car's live position as a live reference to the real world! Many `Cars` in one environment can make up a simple traffic application and can provide some clues to the current traffic situation.

The traffic application cannot "see" the actual implementation. It relies on its interfaces and can not distinguish, if a `Car` is actually making use of a live GPS signal or if it is a computer simulation of a car. The application would have to add plausibility tests to check, if the depicted path is more or less realistic. We first check the car's path against a satellite picture to see if the car is moving on a real road. We then check if the speed of the car is reasonable..., and so on. A car is real as soon as *every* detail of a car and its behavior is represented in the software object `Car`. A `Car` gets more realistic with every detail added to the class `Car`.

LEARNING THE JAVA LANGUAGE > INTERFACES AND INHERITANCE > INTERFACES

> ...Imagine a futuristic society where computer-controlled robotic cars transport passengers through city streets without a human operator. Automobile manufacturers write software (Java, of course) that operates the automobile--stop, start, accelerate, turn left, and so forth.
>
> Another industrial group, electronic-guidance instrument manufacturers, make computer systems that receive global positioning satellite (GPS) position data and wireless transmission of traffic conditions and use that information to drive the car.
>
> The auto manufacturers must publish an industry-standard interface that spells out in detail what methods can be invoked to make the car move (any car, from any manufacturer). The guidance manufacturers can then write software that invokes the methods described in the interface to command the car. Neither industrial group needs to know how the other group's software is implemented. In fact, each group considers its software highly proprietary and reserves the right to modify it at any time, as long as it continues to adhere to the published interface.

Our approach is comparable to scientific research. No matter what kind of science or what kind of experiment, all scientists are searching for more details for the *one reality*.

This book is about objects, things in the real world and since science describes the real world to a very high level of detail it would be a shame not to make use of this knowledge—available online. By "filling" independent objects with scientific facts, they become more real with every new fact added. One can even consider the idea of providing common classes implementing scientific knowledge to be used in any context.

1.4   Models in Physics

Object entities abstract internal processing from the external application logic to simplify the modeling of business logic, rules or algorithms that handle the exchange of information between a database and user interface. The object-oriented approach adds semantics to programming languages and as such, computer language, mathematical language, and human language might be considered to have moved closer together.[2]

> JAVA WHITE PAPER > CHAPTER 3 > JAVA IS OBJECT ORIENTED
>
> ...object technology is a collection of analysis, design, and programming methodologies that focuses design on *modeling* characteristics and behavior of *objects* in the *real world* ...objects are software programming models. In your everyday life, youre surrounded by objects: cars, coffee machines, ducks, trees....
>
> An objects behavior is defined by its methods... you can build entire networks and webs of objects that pass messages between them to change state. This programming technique is one of the best ways to create models and simulations of complex *real-world systems*.

Nevertheless, modeling was not an invention of the software industry. In this chapter, we have looked at electrical devices and cars represented as individual software entities. Since software, devices, and machines are man-made constructs, these analogies might raise the impression that modeling is restricted to engineering. Creating a device and representing the plan in a software language can be part of one process. This section is meant to point out that modeling has evolved from reverse engineering the world. For centuries physicists have created their model of the world. Modeling of real-world objects goes way beyond controlling electrical devices. From a scientific point of view, everything (matter) is composed of a sizeable number of elements and held together by only four forces.

---

[2]You can find the JAVA WHITE PAPER online at java.sun.com/docs/white/langenv/.

What does *modeling of real-world objects* actually mean? What is *real*?

The "methods" of physical objects can be expressed in mathematical formulas. Just like the scientist, the object-oriented programmer is using techniques to *abstract* details to understand the behavior of objects, or things. Reproducible behavior indicates a *physical law*. Physics describe real-world behavior in the language of mathematics. Formulas describe *how* objects behave. These fundamental laws are independent of programming languages and chosen units (i.e., miles or kilometers).

Why not transfer the entire scientific knowledge into software objects? If the idea of software objects is derived from real-world objects, wouldn't it be the ultimate test of object-oriented software to build a *real-world simulator*? Can software objects be programmed to behave like physical objects?

This all sounds like a nice blueprint to create some objects and fill them with mathematical formulas. The testing could be done by comparing simulations with the real world. And in the end: How can you tell the difference between the virtual and the real world? By finding the difference! Mother nature as the project manager should help to minimize discussions!

1.4.1   Object-Oriented Simulations

book.intro.MassObject

This book will not deal with physical formulas. Nevertheless, the idea to build a real-world simulator cannot ignore physical formulas that actually describe the real world in terms of mathematics. In this section, we take a little excursion into a discussion of physical laws, i.e., the law of gravity and how it can be applied to create a tiny real-world simulation with a single class. This class `roaf.book.intro.MassObject` can be downloaded at `www.roaf.de`, where you can also read about how mathematical equations are derived from physical laws and finally implemented into a software simulation. This section superficially illustrates the relationship between physical concepts and object-oriented programming.

Gravity describes the *force* attracting masses to each other, and the formula only requires the *mass* of the participating objects and the *distances* between them to calculate the *acceleration* in space.

Please follow the download and installation instructions in the Appendix to install the `roaf.book` packages, which describe the code used in the first ten chapters. Once you have the packages installed, please execute the method `roaf.book.intro.MassObject.main`.

With two `MassObjects`[3] coded with the laws of gravity and force, creating two mass objects is pretty straightforward:

```
class MassObject { double mass, position; }
```

For programmers familiar with procedural programming, this `MassObject` is like a record set storing the mass' attributes. The programming environment, where the objects are created for the actual simulation, is given by the `main(...)` method. The two participating objects are created by

```
MassObject earth = new MassObject(5.98e24, 0 );
MassObject  moon = new MassObject(7.35e22, 384404 );
```

Physics teaches that all masses are aware of each other—every mass is interacting with every other mass in the universe continuously. In the simulation, the mass objects, earth and moon, have to be introduced to each other:

```
.interactWith(MassObject mass)
```

At this point the `MassObject` becomes more than a record set. Any number of objects can be instantiated from one class, and their methods are familiar with other objects created from the same class. This fact is used in object-oriented programming to transfer or *curry* a two-argument function `distance(moon, earth)` known from procedural languages into a one-argument function, `moon.distanceTo(earth)`.

Basically the method `interactWith` introduces the other mass to interact with and acts as the kickoff *event* to start the interaction:

```
earth.interactWith( moon  );
moon.interactWith( earth );
```

As in the real world, where earth and moon are aware of each other, the simulated objects are connected via *object references*. Both objects are able to inquire the other's *state* (i.e., position) and use it to modify its own state (i.e., current speed) by using `distanceTo()`. The simulation does not give a clue where mass comes from, but the program environment serves as the universe *into* which masses are created to interact. The `MassObject` shows the world from each object's perspective. Following the physical model, force is not a mass attribute and is therefore not represented as a class member!

---

[3]You should follow the code listings as you read.

Physical formulas and mathematical language. It should also be noted that physical formulas can be subject to mathematical transformations. Instead of calculating every moment of the interaction, higher mathematics can provide a formula to predict object states for every moment of interest. By adding a dimension to each mass body, the time of a collision c can be calculated and passed to an `Event`. The basic idea is to pass the information of the collision at time t_c to a timer waiting until t_c occurs, and then fire an `Event` and inform other objects via a `Listener` to give them a chance to react. Events and event listeners (cause and effect) can keep a system running without any external interactions since the objects communicate with each other.

Looking for the function of force in space $F(x, y, z)$ in many different contexts is one of the main tasks in physics. Physically speaking, the gravity experiment takes place in a *closed system* concentrating on the given quantities. By analyzing closed systems, *laws of conservation* can be identified to *predict* results of the interactions. The earth-moon simulation can also be solved with the assumption that the total momentum of the system is conserved.

One of the most powerful concepts is the idea of energy. Energy in physics is like money in economy. While money can be traded to acquire things, energy is a theoretical value calculated in one scenario to be consumed in another one. Nuclear power is transformed via heat to mechanical motion to produce electricity. The total energy of a closed system is always conserved and thus, equations can describe the transformations only inside the system.

In the earth-moon scenario the potential energy of gravity is transformed into kinetic energy of each mass body. By increasing the speed the body is "filled" with energy, and this energy can then potentially be transformed into another form of energy. Although energy is a mathematical construct, it's an intuitive term in our language.

Chapter 2

The Vision of a Real-World Simulator

2.1 Introduction

Chapter 1 compared the functionality of objects to devices, machines, and other physical entities. With the separation of application and object development, we derive a vision in this chapter to develop real-world objects and to place them in (various) real-world applications in order to improve their flexibility and robustness.

The vision will provide a guideline for the remainder of the book and beyond. The goal is to reverse the analogy of software objects and physical objects by implementing real-world objects as independent "things" and to optimize them for different environments. In other words, we put the object-oriented (OO) promises to the ultimate test by developing virtual objects. A combination of different real objects should work without restrictions and reflect the real world in a virtual reality.

2.2 Physical Objects

The little gravity application described in Section 1.4.1 creates masses to interact with each other. Its `main` function represents the *external* environment—the gravity field and a coordinate system to relate the masses to each other. The objects behave according to their *internal* mathematical world model, based on the provided or perceived external information. External and internal development represent two different perspectives on the same world. Simulating the concept of inertia, a `MassObject` requires the attributes `mass`, `position`, and `speed`.

For a physicist a model is realistic if it makes the correct predictions according to the real-world behavior perceived in experiments. Different `MassObjects` can implement different mathematical operations and still interact with one other. To allow for experimentation in simulations, it is helpful to separate the environment from the objects. By checking on the

15

objects' positions frequently, the main program can validate the objects behavior externally.

Another *development cycle* could define the `MassObject` as a billiard ball. This would require the implementation of a shape (ball), size (radius), the inner structure needed to calculate collisions (inelastic) with other balls, and the surface structure to determine the friction of the ball (spin) on the tablecloth, etc.

The idea, however, is that we separate the programming objects and the environments.

JAVA WHITE PAPER > CHAPTER 1 > INTRODUCTION TO JAVA TECHNOLOGY

The Java programming language is designed to meet the challenges of application development in the context of heterogeneous, network-wide distributed environments. ... secure delivery of applications that ... can be extended dynamically. ... to develop advanced software for a wide variety of network *devices* and embedded systems. The goal was to develop a small, reliable, portable, distributed, *real-time* operating platform.

Thus, using Java, one programmer should be able to create and run an environment on his own machine, while another programmer can create and instantiate an object on his computer and hook it up to the environment to interact with other objects of other programmers.

Using the gravity example, the `main` method could represent a "server," acting as the third party, i.e., the ether between the objects. The server provides time and space to detect coincidences and decouple the individual implementations running on separate CPUs and connected via a network. The server could define the "rules" for an object's participation and with the rules, the *degree of reality*. With well-designed physical objects, programmers can run experiments and use predesigned objects to understand physical laws.

In a client-server constellation, the object developer can work in a similar way. He can initially implement a simple world model based on the object's perception (interface/sensors) of the external world. He can then experiment by testing different approaches, starting in very general environments and then refining to special cases.

2.3 Projecting Reality into Virtual Worlds

Theoretical scientists make assumptions about how something behaves and draw conclusions using formulas that are able to predict the results of experiments. This is called a top-down *deductive* method. In simulations, this method is comparable to an object's *implementation* of a mathematical model. On the other hand, experimental scientists use a bottom-up *inductive* method to derive a law. After dropping all kinds of objects from different heights and determining the duration of the fall, the scientist can create a table of masses and durations and look for a formula to predict every value in the table.

In order to keep the vision simple and avoid complex mathematics, the model of a car is more intuitive than the gravity example. Looking at the `Car` object, introduced on page 9, an initial implementation requires only a `position` and some rules for motion, for example, `accelerate` (increase speed), `break` (slow down and stop), and `steer` (change direction). These rules can be validated using a map (a rectangle moving on a line). Then, the next implementation step might be a simple navigation system (see page 103) to guide the car to a destination.

The experimental approach requires real-world data of a car. A `GPS` `trace` (see page 39) can be used to validate the simulation of a car. The acceleration values imply mass and force (engine power), and the track's positions imply a street that should be represented on a map (see page 79).

By combining simulated cars and traced real cars in one scenario, one can validate the map and the behavior of each of the car types. Step-by-step, the scenario and each of the participating cars becomes more realistic. Think of a (city) scenario for busses and another one for subways, which can be projected on a city map as two separate layers. The server provides date and time for each scenario, and the user (observer) can see how a bus and a subway meet at the same station.

Many servers, as different aspects of the same world, could be combined to create one reality! As long as they are reflecting the real world, they should allow projections (layering) of any real-world, real-time information into the simulation.

2.4 Real-World Objects (ROs)

In Java, everything is built on top of the `Object` class to manage the communication of the Java program with the external platform and network via JVM. To build real-world simulations, it would be nice to have a toolkit containing basic classes representing *real-world objects*. Classes of `Cars`, for example, would be useful for simulating *traffic*. Or, a programmer could `extend` a `Car` class to simulate varying set-ups for races.

Definition 2.1
Initial RO Vision

The root class of any real-world object will be envisioned by the `RealObject` (RO) representing any physical thing known from the real world.

Every `RealObject` *must have*

- a *location* and speed relative to the external coordinate system at any *time*,

- a *mass* and a body (*size* or shape) claiming space at its location.

Technically, a `RealObject` should be an independent program to be run independently on its own JVM (and CPU).

As an example, think of a `RealObject` as a smartphone with a built-in GPS receiver and a JVM. The device can be seen as a purely technical (abstract) medium used to collect and transmit information from any real object to which it is attached. The phone number is the reference to its owner. A server application can dial the phone number, connect to a client application, and request its position continuously. The trace would give clues about the owner of the phone, perhaps walking, riding a bike, driving a car, or traveling on a train. By projecting the trace on a map, the difference between a boat and a train ride could be distinguished easily.

This vision is already reality for many objects. More and more products are identified using a unique RFID (radio frequency identification) chip. The range of IP addresses currently allows about four billion computers to be addressed and there are developments to establish new address ranges (IPv6) to supply *every device* with a unique address.

Our vision does not mention *any* implementation details. A `RealObject` could be implemented as a software simulation or as a real-world set-up collecting real-world data live. The object's external reference would not give any clue to distinguish between the two modes. An external observer could only distinguish between them, if the simulation is not realistic.

To clarify the word "real" for a software project, we can modify the Turing test and apply it to project development.

ROAF Turing
Test

Any process, event or object inside a simulation program can be considered to be *real*, if it can not be distinguished from the real thing it is simulating. It is "real," if it replicates something in the physical world without an apparent difference.

Measuring realistic behavior is what the project in this book is all about. (How) can you distinguish two GPS traces, when one is received live from a moving car and the other from a car simulation? It is all a matter of the degree of reality required and the similarity of the behavior to the real world.

Similar to the fundamental Java `Object`, other more sophisticated and specialized objects can be extended from `RealObject`. A `RealCar` might add many attributes to specialize the `RealObject` on the domain car. Step by step, the car can be enriched by an engine, a value for the current tank reading, etc.

2.5   Real-Object Applications (ROApps)

We defined `RealObjects` to specify behavior and the internal coding of physical objects. The RO can be associated with a client program running on a PC. By connecting to a server and logging on, the RO becomes part of a *real-object application* or ROApp. Participating ROs can interact through a ROApp, while the ROApp serves as the third party to propagate events between ROs, observe the RO's behavior, and disqualify them for misbehavior.

A real-object application (ROApp) serves as an external environment for `RealObjects`. In order to enter the environment ROs have to conform to interfaces. The minimum requirement for all RO–ROApp combinations is space and time. The ROApp provides universal coordinates and time for all ROs, while ROs have to be able to report their location at any time. The ROApp serves as a referee for all ROs and propagates information between ROs.

Definition 2.2
Initial RoApp
vision

Technically an RO is a client application and the ROApp is a server application.

This vision does not mention concrete implementations of authentication, authorization, etc. For example, a ROApp could provide a race track and allow `RealCars` to log on and race against one another. The ROApp would be in control to start the race after having approved each car. During the race, the ROApp would propagate collisions between cars and exclude cars for violating racing rules.

ROApp laws are breakable and should therefore be referred to as *rules*, as opposed to laws, such as physical laws, which a `RealObject` obeys. Naturally each client can use any available ROApp information in its internal implementation, for example to guide a car around a traffic jam in the ROApp scenario.

2.6   Real-Object Application Framework (ROAF)

While a RO can be generalized as an actor, the ROApp represents the scenario to act and interact in. Each RO–ROApp combination defines a special domain, a specified area, and a set of rules.

Imagine looking at a city map where you can watch the movement of buses. This ROApp could be realized by simply providing the time schedules for each vehicle, which report the vehicle's location when requested by the ROApp. The vehicles, being ROs, wouldn't have any degree of freedom since the bus driver has to follow a schedule.

One might also consider providing a ROApp for the city's subway system. The city map can be reused, although the subways would follow their own path rather than the street map from the previous city map. Both ROApps are pretty simple and partially overlap.

The next step would be to merge the two ROApps. Merging public buses with public subways sounds pretty simple, but the more you think about it the more rules you'll need.

Definition 2.3
Initial ROAF
vision

A real-object application framework is defined by the laws common to all ROApps, which should be coded in a `roaf` library. Different ROApps (responsible for their ROs) can be overlaid, if they (partially) conform to the ROAF standard. Many ROApps can be combined in one real-world simulation.

Technically a ROAF is a kind of a server farm (or cloud).

2.7   Science Fiction

The ROAF vision was derived from the object-oriented paradigm in software development. This might be feasible for a programming class, but it is not sufficient to attract a user to the project. The word "vision" probably raises a much more exciting expectation, as for example in the context of science fiction. Of course, a science-fiction movie extrapolates from current technology to maintain credibility. So, before you put the book aside here's another way to look at the ROAF vision—simply replace ROAF with "Matrix" and think of the movie![1]

The film poses the question "What is real?" Is everything we sense real or is this perceived reality only created by electrical signals in the brain? We will not discuss how to hard-wire our senses to a computer, although this may become reality one day. The more interesting aspect of *The*

---

[1] *The Matrix* by Laurence and Andrew Wachowski, 1999.

Matrix is the idea of constructing a virtual world to reflect the real world. This virtual world can be entered and exited with given conditions (fixed phone lines). Once the RO client (in *Matrix* jargon, residual self image) has entered, he can interact with the underlying laws governing other objects, for example, gravity. The mission is to discern the difference between the virtual and real world—the ROAF Turing Test. That is, to determine who of the other people is also remotely connected to the matrix, who is an agent, or who is a sentient program only existing inside this virtual world and fighting intruders?

This analogy of the ROAF and the movie, *The Matrix*, should inspire a more lively discussion of the vision of this book—developing the fundamental architecture for a real-world simulation.

Conclusion

You should hopefully now know what the vision is all about? Are three definitions sufficient to clearly define software, and can you create software to realize the vision?

Like most visions, this one is a little fuzzy and it requires effort and experience to transform it into software. Although the vision was derived from objects and physical laws, every reader probably has his own idea of what the RO, ROApps, and ROAF will actually look like. Therefore, the vision has to be formalized to a more technical level.

Before starting any team development effort, the vision has to be transformed into a concrete mission. To minimize time and costs, standards have (been) developed to optimize this phase:

Object-oriented Analysis (OOA) and Design (OOD) Methodologies

Object orientation is key for transforming vague visions into isolated working steps for programmers. The object-oriented paradigm allows software architects to design small software pieces, which can be developed independently in incremental cycles. Central to all methods and tools is the idea that any system is composed of objects with individual behavior (implementation) and those objects are able to communicate with other objects by exchanging information (messages). Experienced project managers can use high-level (project) tools to visually model software and directly generate the source code (skeleton).

Even if the implementation varies, the cognitive process should be similar. The book describes this process in detail, implemented with the Standard Java SDK. The reader can, of course, choose his own implementation.

Chapter 3

OOA—Analysis and Mission

3.1 Introduction

We have combined the object-oriented approach and the scientific approach and developed three simple descriptions or rather expectations: RO, ROApp, and ROAF. Now we need to get a better grip on the system. Architects and project managers usually start their work with formalizing the vision into technical terminology and context. Restricting each person to a specific role and a well-defined task makes work on the project more effective.

The *analysis* phase describes *what* the system should do, establishing a common view and vocabulary for the problem space. This is often referred to as "business logic" (the common language and business between customer and software developers.) The *design* phase describes *how* this can be achieved with various components. This chapter will use a number of language tools to support a common understanding, before starting the actual implementations in Part II.

3.2 Language Analysis

Object-oriented analysis (OOA) is a semantic approach to model a system made up of interacting objects. Just as language is the natural way to communicate in various situations, objects are a key part of the communication in software systems.

Designing programs with meaning (semantics) is the key element of the OO paradigm. Therefore, it is worthwhile to do some language analysis to gain a better understanding of the process. If we use this book as an example, the project manager would receive Part I, "Introducing the Vision." He would then use semantic analysis to break it down into key phrases, highlighting the key words in each chapter:

1. Software Objects and Real-World Representations. The analogy of software objects and devices was drawn with a video application similar to

the software of a multimedia center. By moving the program environment into a remote control, the program flow is physically separated from the participating components, which are only referenced via infrared signals. Each of the technical components, like the receiver, has its own built-in functionality, implementation or behavior.

A real car is another example of a technical device, which can be controlled (driven) *internally*. It can be equipped with a GPS receiver and a cell phone to transmit its position as a *reference* to *external* observers. In car races, *each* team could monitor the race with *common* software, while remotely tuning *their* own car.

The ROAF vision zooms out a lot further by assuming *every* car driving on the globe's surface could be referenced by external applications. Vice versa, this would allow every car to analyze traffic situation for optimized navigation around jams.

2. Models in Physics. What if, every thing, every real object could be referenced by one gigantic framework?

Then, many experiments could be replaced by computer simulations, and the framework would represent (reflect) one reality for all participating objects.

Since the location of a car is not sufficient to ensure it *actually is* a real car, the framework can use scientific laws to validate whether the car is moving according to physical models, described by mathematical formulas. By using additional real-world knowledge, like digital maps as a coordinate grid for cars, plausibility can be verified to a high degree. To make math physical, real-world units have to be introduced. The human environment can be described pretty well using the MKS system with the units seconds for time, kilograms for mass, and meters for distance.

All real-world objects taken together make up one real-world simulation.

3. The Vision. Every `RealObject` must have a weight (mass), a size (dimension), and a location in order to participate in a real-world scenario. The `RealObject` is client software, which can run on its own CPU and use any external real-world knowledge to navigate through a given scenario, a given reality. The server represents a real-object application (ROApp) and serves as a third party to observe the objects' behavior according to a given rule set.

Each `RealObject`

1. *can* have a reference to a real object with a `GPSunit`, a cell phone (transmitter) and a programming environment;
2. can simply play back a `GPStrace` to move through the real world of the server scenario;
3. or can implement a simulation in order to behave according to the perceived reality of the server defined by interfaces.

The *internal* RO implementation *creates* objects, while the *external* ROApp is *referencing* these objects.

Since every ROApp can represent different aspects of one reality, the real-object application framework (ROAF) can combine and synthesize different ROApps into one real-world simulation. While each ROApp is free to define its own (fictional) rules, the ROAF is based on strict unbreakable (scientific) laws; i.e., drivers can break speed limits, but they can not overcome the force (inertia) pulling their car out of a curve.

Reality is defined by a network of real objects exchanging information.

3.3   Semantic Network

The highlighted keywords can be separated into technical terms, like *programming environment* or *application*, which are not part of the vision. The remaining vocabulary makes up the content of a semantic network.

This early stage of project management is time consuming, but it is also a very useful investment. In addition to the technical advantages the Internet provides, it is a semantic tool and it is worthwhile to search for linguistic tools. There a numerous tools available that accept two expressions and return a semantic diagram connecting them. Using one of these tools, you might end up with a semantic network like the one in Table 3.1.

With the help of a semantic net, you can discuss the project vision from left to right, or top to bottom, and find out, if everything has been covered. Language analysis and semantic networks are part of a modeling process to support the project manager's thinking, but they are usually not part of the project documentation used by the developer team.

The project manager may now come to the conclusion that "real-world simulator" is a good working title and expresses more than RO, ROApp, and ROAF, which refer more to different technical modules. These relationships can be depicted as in Figure 3.1

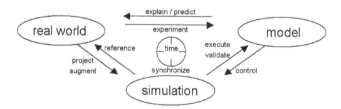

Figure 3.1. The semantic analysis supplies a rough idea of the purpose of a *real-object application framework*. The framework combines a virtual world model with augmented real-world information to produce a simulation.

SCIENCE   facts   PHYSICS   interactions   MODELING
      universal   platonic   indicates   assumptions
discoveries   METEOROLOGY   equations   THEORY   scale
      weather   temperature   prove   abstract   dimensions
2D   measure   units   law   where?   MATHEMATICS
      GEOGRAPHY   experiment   verification   logic   statistics
maps   routes   variation   COMPUTERS   stochastic
      TECHNOLOGY   GPS   tests   validation   numerical
devices   coordinates   locate   where?   programs   operations
      machines   trace   algorithms   operators
GEOLOGY   REAL WORLD   explain   REAL-WORLD MODEL
      entity   environment   exact   KNOWLEDGE
GIS   3D   NATURE   prediction   evaluation
            force   earth
balance   gravity   evolution   control   RESOURCES
      weight   MATTER   MKS   relational   persistence
atmospheric   mass   SPACE   synchronize   data   structures
      BIOLOGY   infinitely   TIME   live
body   organic   how?   coincidence   real-time   fiction
      perception   LIFE   detectors   game   libraries
anthropology   HUMAN   augmented   projection   provide
      behavior   REAL-WORLD SIMULATION   software
communication   believe   distributed   encyclopedia   internet
      cognitive   sense   virtual   realistic   INFORMATION
sociology   interpretation   test   probability   literature
      PSYCHOLOGY   word   robotics   artificial
implications   meaning   classify   INTELLIGENCE   agents
      grammar   syntax   network   neural-network
semantics   LINGUISTICS   competitive   brain
      translate   sentence

Table 3.1. Semantic net for a real-world model simulation.

A `RealObject` Generator

Wouldn't it be great, if the process from language analysis to semantic net
could be automated? The next generation of programming tools might be
able to look up words connected via semantic allocations and create code
shells with implicit domain knowledge. Today's online computer represents a system with an infinitely large database of the entirety of human

knowledge. The challenge (and risk) is to identify and give meaning to the information.

Before starting the development of a `RealCar`, for example, search engines could find expressions and objects used around a car. If language and knowledge are needed to create `RealObjects`, it would be desirable to automate a preselection process with a `RealObjectGenerator<RealCar>`— software that would collect all scientific facts about cars and their semantic context (as connectors to other objects).

3.4   Gathering Information

The semantic net is supports creative thinking, but project vocabulary is not project knowledge.

Object orientation suggests to gather domain knowledge, in order to design objects similar to real ones. In addition to the gravitational formula, it helps to know that gravity is tied to every mass. Knowing this, the architect has to make sure that the gravity law is always applied, wherever the mass attribute is added to an object.

The first topic (see page 35) we will need to gather material for will be space and time, the prerequisite to describe motion (= space/time). The globe has a coordinate system to find any place on its surface. However, spatial coordinates are not very practical to describe physical laws of motion. These are usually expressed in linear Cartesian coordinates. In this case, the main task of gathering material is to answer the question: How can global positions be converted into linear Cartesian coordinates?

Gathering information is not restricted to only the beginning of a project. For new software development where large user groups are involved, it is often worth the effort to conduct interviews or send out questionnaires on a regular basis. Not only does the software itself evolve, but also the perception of usability can change over time and the users, not the developers, are the experts.

3.5   Data Dictionary

The semantic net is a useful exercise, but still leaves room for interpretation. Like an appendix or a glossary, project papers should be based on a *data dictionary*, which should be maintained regularly to reflect current project status. A great maintenance tool to prevent confusion and improve communication is a wiki software accessible to the entire development team.

A data dictionary should establish and define the major vocabulary of a project in a common context. Since this project is about the vision of a real-world simulation, one might start with an entry for the term *simulation*.

simulation

> A research and learning technique to reproduce *real* processes initi-
> ated from *real* events in the *real* world. Simulations execute math-
> ematical models of physical systems on a computer. The computer
> is used to represent the dynamic responses of the real world, by the
> behavior of the software simulation modeled after it.

In addition to the initial vision of RO, ROApp, and ROAF, the ROAF
Turing Test, that was described on page 18 in order to find a practical way
to actually test the word real or to measure a degree of reality should be
in the data dictionary.

ROAF Turing Test

> Any process, event or object inside a simulation program can be
> considered to be *real*, if it can not be distinguished from the real
> thing it is simulating. It is realistic, if it replicates something in the
> physical world without a difference.

All ROs and ROApps rely on the degree of reality of the other and
should validate any information against the real world whenever possible.

As you continue to think about and discuss a project, you will discover
more words to add to the data dictionary. More terms are included here to
provide the idea of the process; the reader can make his own list or discuss
this one:

real object

> Any physical object, with a mass, a size (body), and a geographic
> location on the globe. *All* physical objects are *ruled by the laws* of
> classical physics.

real world

> The sum of all *real objects* and their interactions (events and pro-
> cesses).

interactions

> Real objects interact by referencing (perceiving), invoking methods
> on each other and reacting (processing). In the physicists real-world
> model, objects interact by exchanging particles. Simple interactions
> like inelastic collisions can be described and calculated with conser-
> vation laws. People can interact via communication by exchanging
> words (strings).

model

> A mathematical representation of an object, event or process. A
> single model may contain multiple integrated models, each of which
> represents a level of *granularity* for the *physical system*.

roles

> To understand reality, artificial objects can be built and used to dynamically act out *roles*. Each object acting a particular role perceives reality from a different perspective.

intelligence

> Processing information and making a choice is the smallest unit of intelligence. Intelligence will not be provided by the framework and has to be implemented by the programmer.

3.6 Problem Statement

After the initial analysis phase, with the project vocabulary and (basic) domain knowledge established, the project manager transforms the vision into a software product. Modern software development evolves in life cycles. Some experience is required to define the first product version and divide it into highly independent components, connected only by well-defined interfaces—the basis for parallel teamwork. Teams should talk about interfaces (contracts) and not *how* they are implemented. Each component can be developed individually in its own iterative, incremental cycle to be predictable, reproducible, testable, and replaceable. This *evolutionary* structure improves the product quality and makes the development controllable.

Although the problem statement should describe the final product, a project manager can identify the vital building blocks and abstract them to a level, that allows early implementations. One advantage of software objects is their usability with very few implementations required to produce a first product (or demo) version. Then, every developer can code using running software and gets immediate feedback if something doesn't work.

This book will focus on an initial minimum version of a real-world simulator to demonstrate the basic functionalities, allowing the reader a basis for future development and exploration.

ROAF problem statement. The ROAF is a system to simulate realistic environments (ROApps) for physical objects (ROs) and vice versa (complementary): The ROAF is a system to project real objects into a simulated environment. The system can validate the objects' behaviors with scientific models to coordinate them in time and space. Theoretically, the ROAF could contain *all* aspects of reality. Practically speaking, it is a toolkit to allow the development of isolated ROApps to simulate only certain aspects of reality. ROs and ROApps are adjusted to each other by well-defined interfaces of the ROAF.

The ROAF is a system—or more precisely, it's a framework, a kind of library.

What is a Framework?

In ROAF terminology, a framework supplies laws for yet unspecified applications. A developer should be free to create his own ROApp, while conforming to general standards in order to be compatible with other ROApps. Technically speaking, the ROAF is a set of classes providing technology and solutions to code real-world simulations. This common basis afforded by a real-object application *framework* reduces programming effort, and helps to enforce consistency and to increase interoperability.

Later, in the implementation phase, we will see how the idea of a framework functions by looking at the Collection framework to collect spatial coordinates and the Swing framework for mapping spatial coordinates on a Cartesian grid. The ROAF observes ROs and propagates information among them.

Using object-oriented programming, a framework can be coded with a minimum implementation to get the system up and running before adding more details in further iterations. The ROAF can be coded around the `RealObject` with minimal common attributes of every physical object. GPS attributes are sufficient to trace an object and the framework can implement the information exchange technology, for example.

Besides being a framework, the ROAF has the potential to create libraries in which things can be plugged in. The `RealObject` is only specified to be any physical object. Just like a library full of books on different topics, the `RealObject` can be extended to become any thing. A developer might choose to program `RealCars` and publish this library of different cars to be used inside the ROAF. An application developer for a race track could rely on the compatibility of the cars. In object-oriented terms, any subclass of an RO for any problem domain should function in the framework.

3.7   Candidate Objects

The process from vision to language analysis to a problem statement does not require programming skills—the exact intention of OOA. Thus, the customer for a particular software product can and should (for your own sake—he must!) participate in the initial analysis phase.

The ROAF project was derived from the usual introduction of software objects and, therefore, the candidate objects are easy to identify. In a traffic simulation the central object is the car, while GPS receiver and sender are supporting objects. In project terms, a `RealCar` *is a* `RealObject`. A `RealObject` *has a* `GPSunit` indicating its `Position` in spatial coordinates.

3.8 The Mission

The end of the analysis phase marks the beginning of the programming project. This book only indicates how to apply OOA methodologies and the reader is advised to discuss these ideas with other programmers in order to develop their own visions of the framework.

The most important aspect of the mission is to define the end of a project—to have a well-defined goal. Without an actual implementation, the definition of the end may be fuzzy. The phase from the first lines of code to a running demonstration platform is the mission of this book. We describe the process in order to hopefully establish a development life cycle for an online community at `www.roaf.de`.

A technique for rapid development might be described as "minimum implementation with maximum abstraction." Object-oriented development enables architects to work on different levels of granularity, similar to filling a large hole. At first big rocks are used to roughly fit the shape, then smaller rocks fill the space between the big ones, and finally fine sand is added to fill remaining gaps.

On page 9, a car was introduced as an object participating in different scenarios with other cars. This simple setting can easily serve as a sample scenario: A car (or vehicle) is a special type of RO, many cars make up traffic, and the ROApp scenario is a given area (i.e., city) on the globe.

By defining a reference application (or presentation prototype) the project manager defines a goal—together with the software customer. The customer is probably not interested in whether the software is object oriented. He can perhaps understand that different cars can be simulated on client machines, while they are represented on a server frontend—a map. From here on, the customer's view (cars, traffic, city) are decoupled from the programmer's perspective (RO client, ROApp server).

Of course, the scenario is still vague, but for programmers it is much more useful than the initial vision. Programmers can begin concrete coding without additional details.

The mission of this book is

- to create a distributed client-server application with
    - the clients representing vehicles (players)...

- to populate a server application (game observer) with
    - a map scenario (game board) and a ruleset (controller).

In other words, the book will provide a prototype `roaf v1.0`, which marks the end of the initial coding from scratch and the beginning of an evolutionary life-cycle development. The prototype makes communication between all parties *much* easier, including product definition, customer presentations, and marketing research.

After having read our initial vision, you may have developed different scenarios. You should keep it in mind! OO propagates the development in life cycles to avoid major redesign issues. A prototype is simply the start of a new cycle, and after reading the book, you can use the prototype as a template for your own scenario.

Part II

Global Positioning

Now that the vision has been thoroughly discussed and analyzed, it is time to begin the mission. The vision was split into three main components RO (client), ROApp (server), and ROAF. Before tackling the implementation of these components, this part will provide the prerequisites to synchronize real-world data in space and time.

Physical laws of motion are generally described in isolated values, beginning at the (x, y, z)-coordinate $(0, 0, 0)$ and moving along the x-axis. These laboratory coordinates neglect the relative where and when. In order to relate real-world activities, every event has to be translated into a global coordinate system and absolute time. Motion has to be transformed from spatial coordinates to metric units to reflect, for example, the speed of a car on a map. GPS coordinates and projection systems will be discussed in order to build a virtual GPS unit as the heart of the ROAF to synchronize object motion in space and time.

The three main components of the ROAF do *not* specify a visualization frontend. There are simply too many commercial and free (map) tools on the market for drawing with spatial coordinates. Nevertheless, a very simple `MapPanel` will be developed to visualize positions and routes for ROAF development. This process will also be used to point out the idea of a framework by looking at the architecture of the *Java Swing Framework* with the platonic `Component`. This excursion should help to clarify the vision of the ROApp framework with its platonic `RealObject`.

At the end of this part, the components for global positioning, time keeping, and drawing of spatial data is combined in a GPX viewer. GPX is the most commonly used (exchange) format for GPS data. This tool, with its subcomponents, should come in handy for every ROAF developer to analyze spatial data, which is available on the internet.

Chapter 4

Space & Time

4.1 Introduction

In this chapter the `roaf.gps` package will be created to manage spatial (WGS84) coordinates and to keep universal time (UTC). A virtual GPS unit is created to play (back) or record GPS traces and to provide transformations from geographical to Cartesian coordinates in order to calculate absolute distance and speed.

4.2 A `GeoPoint` for Spatial Coordinate Systems

The RO vision on page 18 states that "...every `RealObject` *must have a location* ..."

A location? Since every ROApp represents a virtual real-world scenario, it makes sense to restrict the real world to the surface of the globe. Some research reveals the information needed to create a general-purpose position object:

- Geophysically speaking the globe has a fluid core and a cold crusty surface.

- Gravity forces its mass to form a ball spinning with one revolution per day.

- Mathematically the spin implies two well-defined points: the poles.

- Due to inertia the poles are closer to the center than the equator (flattening).

This geometric body suggests a spherical coordinate system and finally defines any location on the surface with the `double` values `latitude`, `longitude` for two dimensions. The vision of real objects separates internal coding from the external environment. Therefore the third dimension `elevation` in meters above sea level should be added for three-dimensional aspects. A car driving up hill behaves different than a car rolling downhill due to shifted gravity forces relative to the car. The majority of the globe's surface is at sea level, so the elevation should default to 0 (plane), if not supplied and is better than using a `null` value.

> Please execute the method `roaf.book.gps.GeoPoint.main`.

The class `roaf.util.GeoPoint` is created from scratch similar to the standard Java classes `Point2D` and `Point` with the methods `get/set`, `longitude/latitude`, `distance`, and `move`. The default constructor (no arguments) can not be used in order to enforce concrete values for a position at construction time. Decimal values are used to avoid working with degrees, minutes, and seconds. The methods `latitude` and `longitude` have `private` visibility and can only be accessed by getters and setters to be validated internally to ensure that the values fit into the geographical grid:

$$\text{latitude: } [-90°, 90°], \qquad \text{longitude: } [-180°, 180°].$$

Implementing a `distance()` method turns out to be more complicated than in a Cartesian coordinate system.[1] A spatial coordinate system on the surface of an ellipsoid uses dimensionless angles (relative), and size and shape are needed to calculate absolute units

For a symmetric body, two values—those for mean radius and the flattening of the earth—are sufficient to use the Haversine (half-versed sine) formula, a mathematical equation important in navigation, giving great-circle distances between two points on a sphere from their longitudes and latitudes.[2] The table below shows the values calculated using the formula for the metric distance of one degree in the longitudinal direction:

`roaf.util.GeoPoint`

```
Euclidian distance of one degree longitude
at latitude 0.0 is 111.1864599561611 km
          15.0 is 107.3983296596687 km
          30.0 is  96.2918262451686 km
          45.0 is  78.6231940795413 km
          60.0 is  55.5958756444310 km
          75.0 is  28.7788771173616 km
```

Note that the distance method only accepts one *other* point to calculate the distance from *this* geographical point. The object-oriented programming technique of *currying* transforms a two-argument function `distance(from, to)` into a context sensitive function `this.distance(to)`.

Besides calculating *how far* away another object is on the globe it is vital to find out exactly *where* the object will be located after moving a

---

[1] The Cartesian distance in two dimensions is determined by $d = \sqrt{x^2 + y^2}$.

[2] You can find the formula at **`roaf.util.GeoPoint.distance ( Position )`**.

certain distance in a specified direction. We therefore add a new method, `move( direction, distance )`.

4.3 A `Position` Interface for Various Coordinate Systems

Basically the `GeoPoint` method is now ready to be used to represent any location of any entity on the globe. However, the earth is not a perfect mathematically regular body. Its shape is a rather irregular *geoid*. The hard-coded implementations of the `distance` and `move` method in the `GeoPoint` method are hardly ever precise on the world's surface in every continent, country, or region. This is a typical programming problem. Instead of finding the best compromise for the method's implementations, *interfaces* provide a way to separate *what* and *how?* What is the distance? How is it calculated (on an ellipsoid)? Note that the `roaf.util` package *is not* an essential part of the `roaf` library. It supplies useful, general default implementations out of the box and can easily be replaced. By using an interface with the `GeoPoint` method, the actual implementation is separated from the formulas or method used.

The interface only defines method signatures[3] expected from a class:

roaf.gps.Position

```
public interface Position
{
// public abstract void setLatitude(double latitude);
// public abstract is not needed:
   void setLatitude (double latitude);
   void setLongitude(double longitude);
   double getLatitude ();
   double getLongitude();
   double distance (Position pos);
   double direction(Position pos);
   void move( direction, distance )
      :
}
```

The interface *is* part of the `roaf.gps` package and vital for the `roaf` library. Every position handled inside the library should be coded using the `Position` interface instead of a hard-coded implementation. With this interface every real-object application can use its own formula for distances. This opens up the possibility to link external context-sensitive GPS libraries. One scenario might describe a racing track and require centimeter precision, while another might deal with airplanes connecting the continents and require a quite different measure. A major purpose of libraries (or frameworks) is to separate things that change from things that don't.

---

[3]The method's name and parameters.

> The GPSapplication class and main method were written to demonstrate the usage of every class and interface in the roaf.book.gps package.
>
> Note that the method gpsUnitDemo() requires resources from your hard drive. Please refer to the Appendix for instructions on downloading the resources and adapting the paths to your environment.
>
> It is advised to run the method GPSapplication.main now and make sure the resource GPX file is allocated properly. Then, you can analyze the details of the demo class as you progress through the rest of this chapter.

4.4   A Route to Manage an Array of Positions

A route is an ordered list of geographical locations (lat, lon) and represents an essential tool to handle spatial data in various contexts. Java comes with the *Collection Framework* and provides a number of implementations to collect objects. Our real-world constructs are not included there and must be developed separately.

> Please execute the method roaf.book.gps.GPSapplication.main and study the invoked method routeTraceDemo() as you proceed through Sections 4.4–4.7.   You can chose to comment out the method gpsUnitDemo() for now.

roaf.book.gps.Route

A new class roaf.book.gps.Route is created to hold the private member ArrayList route. Since private members are not externally accessible they are "wrapped" in the class and the developer can add external access to dedicated members of the hidden class. "Wrapping" the route in the class is a good approach and allows you to later unwrap the ArrayList API step-by-step in the development process. This way you can control the visibility of methods as needed.

Although the ArrayList is hidden inside the Route class, the programmer should be aware that a collection generally holds object references. Therefore, a location of a route can be modified without updating the route attributes. To emphasize this fact, the method Position.getNewPosition() is added to the interface. Every location object implementing a position should provide a method to create a new location as a copy:

```
public void appendPosition( Position point )
{
    // encapsulate a new object as new Position
    sequence.add( point.getNewPosition() );
```

```
       expandBoundBox( point );
       calculateTotalDistance()
           :
   }
```

The method `appendPosition` causes a new position to be created explicitly without an external reference. The bounding box and total distance can not be modified by shifting a location externally. The `route` variable and its elements are *encapsulated*, while the actual implementation can remain unknown. Conversely, positions should be copied when retrieving them from the route:

```
   public Position getPosition( int position )
   {
      return sequence.get( position ).getNewPosition();
   }
```

The price of encapsulation are the cost of additional objects. One point is created and a copy is appended to the route. To move a point, yet another copy is created and copied one more time to put it back into the route.

4.5   NAVSTAR GPS

The next question after creating the `Route` and `Position` methods is how to fill a route with plausible real-world data. The obvious solution is a global navigation satellite system (GNSS) or global positioning system (GPS). While GPS is the currently used shortened acronym, it was derived from NAVSTAR GPS (Navigation System for Timing and Ranging), the first actual global positioning system, launched in 1971. About 24 satellites orbit the earth. With at least four satellites (three for position, one for time) in sight, a GPS receiver is able to calculate its (`lat`, `lon`) location with a certain precision and speed, depending on a number of factors. In addition, many GPS devices are able to mark and record (or even transmit) the current position. More sophisticated units allow the user to create and store routes as well as load (digital) maps to be displayed on a small screen. A simple mouse could be considered a GPS representing `Position` in the software world. GPS systems are becoming more and more prevalent and can be found in a growing number of devices (cell phones).

From the abstracted project view, a GPS is simply an arbitrary technical system to enable electronic receivers to determine their current longitude, latitude, and elevation information. Part of our task is to create a GPS receiver for the software world, a *GPS unit*. Every RO will have a built-in *virtual* GPS receiver.

4.5.1   NMEA 0183

Although there is a (growing) number of GPS devices and vendors on the
market, they all perform the same core task—acquire their current position
via satellites. In the early years of GPS device development, a standardized
data transmission protocol called NMEA was defined. The NMEA 0183[4]
format allows a vendor-neutral development of GPS applications. Many
PC map applications allow the user to connect a GPS NMEA device to
the serial port of the computer and display its position on the map. The
NMEA 0183 Interface Standard defines specific sentence formats (ASCII)
for a 4800-baud serial data bus (RS-232-standard for COM ports). Each
sentence begins with a `$`, has a maximum of 82 bytes, and ends with `CR/LF`.
NMEA basically defines twelve sentences to describe different kinds of information.

> You can find a sample NMEA file in the resources at
> `../resources/gps/HD/demo.gpslog`.

The JavaGPS project comes with the small NMEA file `demo.gpslog`.
Although vendor-specific sentences are allowed, each NMEA GPS device
should be able to provide the recommended minimum sentences `GPRMC` and
`GPGGA`, which adds height (and fix) information. The individual sentence
elements are explained in Tables 4.1 and 4.2.

`$GPRMC,134030,A,4924.5644,N,00842.8216,E,0.0,191.3,010302,0.2,E,A*1D␣[CR/LF]␣`

`$GPGGA,134028,4924.5643,N,00842.8212,E,1,07,1.6,200.8,M,47.9,M,,*46␣[CR/LF]␣`

In addition to NMEA log files, the satellites' status information and
graphical front ends are freely available on the internet. Imprecise measurements may be attributed to the topology of satellites or of the actual
location (in a deep valley). At this early stage of our project, we will
assume that every device does the best it can to determine its latitude,
longitude, and elevation; we will, therefore, simply ignore the remaining
NMEA sentences.

4.6   GPSpoint implements GPSinfo

GPS was designed for high-precision detection of latitude and longitude.
Since satellites are more than 20,000 km above the surface of the Earth,
for geometric reasons the height value *can not* be as precise as latitude and
longitude. A barometer can be used to get a more precise determination
of the elevation, and the measurement does not depend on a clear view on

---

[4]National Marine Electronics Association at www.nmea.org.

GP	a GPS receiver
RMC	recommended minimum sentence C
134030	universal time coordinated (UTC) time: 1 pm, 40 minutes and 30 seconds
A	data is okay (no warning)
4924.5644	49 degrees and 24.5644 minutes
N	North
0842.8216	08 degrees and 42.8216 minutes
E	East
0.0	speed (relative to last position and time)
191.3	bearing (0 means headed North)
010302	date 1st of March, 2002 (ddmmyy)
0.2,E	magnetic declination
A	mode (estimated, differential, simulated, ...)
*1D	checksum of this sentence

Table 4.1. The individual elements of the NMEA GPS device GPRMC sentence.

the sky. Thus, some GPS devices have a built-in barometer and compass to determine elevation and course independently.

Note that bearing, distance, and speed are not generic; this data is derived from GPS information since a GPS provides single point information only! From the programmer's point of view, GPS basically adds a time stamp to the GeoPoint and Position information and could be added to their definitions in the source. On the other hand, both definitions of the class and interface fulfill their purpose to localize something. Therefore,

GP	a GPS receiver
GGA	system fix data
134028	universal time coordinated (UTC) time: 1 pm, 40 minutes and 28 seconds
4924.5643	49 degrees and 24.5643 minutes
N	North
0842.8212	08 degrees and 42.8212 minutes
E	East
1	GPS quality indicator
07	number of satellites in view
1.6	horizontal dilution of precision
0.2,E	magnetic declination
200.8	elevation above/below mean sea level
M	elevation in meters
47.9	geoidal separation

Table 4.2. The individual elements of the NMEA GPS device GPGPA sentence.

we leave them unchanged and instead *inheritance* is applied to `extend` the existing classes and interfaces:

```
public interface GPSinfo  extends Position
public class        GPSpoint extends GeoPoint impl. GPSinfo
public class        GPStrace extends Route
```

Inheritance is somewhat the opposite approach to that of a wrapper in that it supplies *all* inherited methods of the inherited class as a starting point. Then, each method can (or has to) be overridden in the development process. Inheritance is often referred to as "*is a*" relation. A `GPSinfo` *is a* `Position` and can directly make use of its methods. In the program flow, a `GPSinfo` can be turned into a `Position` by simply casting it.

A programmer applying the classes of the `roaf.gps` package should understand the technique of casting. Casting does not create or modify objects; rather, it changes the type of a variable into another type as long as the conversion makes sense. Objects are only created with the new operator and remain untouched from casting.

Now, the classes and interfaces `Position`, `GeoPoint`, and `Route` can be used to model geometry on the globe while `GPSpoint` and `GPSinfo` add the time provided by a GPS device.

4.6.1   Universal Time

Global positioning relies on atomic-precision synchronized time and, as such, every GPS device is a universal atomic-precision clock. Just as *global* positioning raised the need for a unified model of the globe, *unified* time was also necessary. It was defined in 1984 as the World Geodetic System (WGS-84) and is based on international atomic time. Coordinated universal time (UTC) can be considered one single clock (instance) running at a well-defined place (the zero meridian).

Every (local) time stamp has a relationship to the UTC clock; A global application has to be able to deal with `timezones` in order to coordinate different events throughout the globe. UTC is sometimes referred to as "GPS time," since this is the time reference of the GPS satellites. The Java Standard Edition provides the `Date` class[5]:

> The class Date represents a specific instant in time, with millisecond precision. ... the Date class is intended to reflect coordinated universal time (UTC) ... the Calendar class should be used to convert between dates and time fields and the DateFormat class should be used to format and parse date strings ... TimeZone represents a time zone offset, and also figures out daylight savings.

---

[5]download.oracle.com/javase/6/docs/api/java/util/Date.html

The tricky thing about the `Date` class is that you can never really "see" the actual date. If you request the current time with `now = new Date()`, it internally sets the UTC time. By printing `System.out.println( now )`, the PC environment does the formatting to its local time zone.

The problem with time zones is that they are actually political borders (and potentially subject to change) and do not simply reflect the path of the sun. Although they could be determined by `lat` and `lon` implicitly, it would require a large look-up table, which is not provided by the Java Standard Edition. Nevertheless, `GPSpoint` uses a `naturalTimeZone` given only by the longitude's offset to the zero meridian, which can be quite useful as a rough estimate.

The ROAF is designed to be a global application and should carefully ignore the local PC time and time zone. This actually means that time and date could be dropped completely, since the objects are supposed to interact directly. On the other hand, date and time will be needed to schedule events.

During the development phase, all applications can run on one computer and implicitly use the same clock; in a distributed application, the server might provide a common time. And, in the long run, all participating machines could be required to synchronize with a network time protocol (NTP) server, while all client devices with a GPS receiver can directly access the server's clock signal.

4.7 GPStrace extends Route

Note that a route *can be* drawn as a continuous line composed of one or more line segments (polyline) on a map. Nevertheless, a route is only a collection of positions, which can be fed into a navigation system to be connected via the streets of a digital map. Different systems can navigate the same route in different ways.

The information stored in a NMEA file define a "trace." A trace is more than a route enriched with times stamps—a trace describes the exact measurement of a motion of an object. Most GPS devices are designed to provide position and time every x seconds or every y meters. In practice, the time span and precision vary, but every point of a trace can be interpolated to reflect the actual motion.

Traces make routes authentic and are very valuable for real-world simulations. By carefully analyzing and validating a trace, a number of assumptions can be made:

- The top speed gives a clue as to whether it was recorded on a bicycle or a car.

- A large acceleration could indicate a sports car or a motorcycle.

- A trace of a pedestrian can reveal secret paths through a neighborhood.

- Traces of one racing car on different days might imply varying conditions (weather, etc.).

- Different traces of an athlete's work-outs might reveal his condition.

While the route is used to describe (a rough) geometry, the trace relies on physics and motion. We create another class `GPStrace extends Route` to finally establish a useful set for very general purposes. The four classes `GeoPoint`, `GPSpoint`, `GPStrace`, and `Route` and two interfaces `Position` and `GPSinfo`[6] are designed with OO techniques and reflect the semantics to use GPS data.

Internally the `Route` is programmed with a `Position` type object, but in an external environment it can hold any object implementing a `Position`. It can mix `GPSpoints` and `GeoPoints` at the cost of type checking for upcasting later. In a distributed application every local environment can work with its own implementation. A `new GeoPoint(...)` can never be upcast to a `GPSpoint`, since it has no time stamp and therefore, the conversion would not make sense.

Internally, the `GPStrace` makes use of the

```
protected List<Position> sequence;
```

by implementing additional bookkeeping of the element's types. Note that this provides *protected* visibility for inheriting classes. Another useful implication of inheritance is the distribution of responsibility. The `GPStrace` is responsible for controlling the time stamps, while relying on the `Route`'s management of the `Position` part.

It seems that a programmer can cast the object hierarchy up and down, but there is one important catch: no matter which "perspective" you chose to look at an object, a method call always invokes the highest overriding method. Thus, if a `GPSpoint` is cast to `GeoPoint`, `GPSinfo`, or `Position`, the method `getNewPosition()` will invoke the `GPSpoint` implementation while the other implementations remain obscured.

The `GPStrace` can make use of the `Route`'s method as any client does. The total distance is managed by the `Route` and the total duration is managed by the `GPStrace`.

4.8   JavaGPS

We now want to create a component making use of the `roaf.book.gps` package and supply space and time information whenever it is needed.

---

[6]See Figure 4.1 for the UML diagram.

For example, an application of our component might be attaching a GPS smartphone to any object in the real world and retrieving its location by calling it.

Before a programmer starts to code a component from scratch, it's generally a good idea to look for existing software and explore how it works and how it might be modified. At SourceForge, there is a pretty good match, written by Ulrich Walther, for the component we want to use as a basis for our code:

`sourceforge.net/projects/javagps` or `javagps.sourceforge.net`

> JavaGPS (V1.02) is a Java-only base library that enables access to GPS devices from within any Java application. Provides Java API, NMEA0183 parser, record and playback GPS log files, convert between earth dates and Gauss Krueger, GPS management GUI with map.

The project given there gives a nice overview, preview, and review of global positioning coding and does not require very high-level Java. You should look at this library and then be able to create your own for the ROAF.

4.9   Requirements for a `GPSunit`

With a universal definition of space and time, all real objects and real-object scenarios can be synchronized and positioned. The `roaf.book.gps` package serves as a toolkit to manage geographical and metric coordinates. We now create a `GPSunit` to keep track of time and space. This virtual device can be added to any object to capture its motion. Every RO has a built in GPS receiver or, in the *simulated* software world, a *virtual* `GPSunit`.

It is always a good idea to set up a requirements or feature list *before* starting to code. For the JavaGPS project, the development steps might include the following considerations:

1. A `GPSunit` determines its `Position(lat, lon, elev)` at any point in time described by the `java.util.Date` class. Waypoints[7] and Routes can be loaded to point the user to his destination.

2. A better `GPSunit` can provide live traces (or positions), record them, and replay them to an application as if they were live.

3. Some GPS devices have a barometer (for elevation) and/or a compass (for direction).

4. A nice feature is a "proximity alert" to inform the user that he is a certain distance from a given position.

---

[7]Waypoint is a GPS term for a `Position(lat, lon)` with a name.

5. Additional features (found in many smartphones) would require more sensors for acceleration and equilibration.

6. A more sophisticated `GPSunit` can display (digital) maps with the trace.

Acquire Time and Place

> Please execute the method `roaf.book.gps.GPSapplication.main` and study the invoked method `gpsUnitDemo()` as you proceed. You can chose to comment out the method `routeTraceDemo()`.

When a GPS device is switched on, it searches for a satellite to acquire the exact time and keeps searching for more satellites to determine its current location. While the first GPS receivers in the early 1980s weighed more than 20 kg and were the size of a suitcase, our `GPSunit` is virtual and represents the location of the RO owning the unit. Time and place are provided by the RO's environment, the ROApp. The `GPSinfo` implementation is vital since it supplies the coordinate system, universal time as well as conversion tools. Here's a start for the new class:

```
public class GPSunit
{
   public GPSunit( GPSinfo   initialGPSinfo)
   { currentGPSinfo = initialGPSinfo.getNewGPSinfo(); }
   private GPSinfo currentGPSinfo;
}
```

The satellite information has to be supplied at startup time. The constructor represents the on switch and initializes the `GPSunit` to current time and location. The semantics indicate that the device starts up with the `initialGPSinfo` to set the `currentGPSinfo`. The idea is to encapsulate the `GPSinfo` implementation as a private member with one single instance during the entire lifetime of the `GPSunit`. Whenever the unit is moved, the instance is moved with `currentGPSinfo.move`.

The time information provided is treated a little differently. The environmental time is defined at the construction time of the unit and is stored as an offset to the time of the machine's operating system. The PC hardware is responsible for keeping track and for changing the time; a real GPS unit doesn't necessarily need an internal clock.

The initial design of a `GPSunit` could look like this:

```
+ GPSunit    (GPSinfo)    set place & time 'switch unit on'
+ setGPSinfo (GPSinfo)  reset place & time
+ getGPSinfo ()           get place & time 'now'
+ setPosition(Position) reset place
+ setGPSdate (Date)     reset time
```

4.10   Real vs. Simulated Motion

As seen above, the `GPSunit` does not really acquire its `Position` and has no chance to validate it. When a real car is moving, it might lose the satellite connection occasionally; the GPS simply recalculates the new position as soon as the satellites are visible again. Built-in navigation systems are able to take the speed and steering direction (telemetry) into account.

The mechanism in the simulated world has to be different, yet provide the same results. The ROs moving around in the ROAF can be real (a moving GPS device), a playback of a trace, or a simulation of motion. An external observer can only validate the trace against reality. This means that the simulation mode has to update the `Position` as it moves. Now, the `GPSunit` goes beyond a real GPS device and is extended by the simulation method:

```
private void simulateMove(double direction, double speed)
```

A real GPS trace serves as a top-down approach, since it describes the results of a measured motion. Although simulation is a more laborious bottom-up approach, it may be preferred over a trace, since a trace does not describe the precise motion between two points, but rather only uses the average speed from the time stamps and distance. The more points the GPS acquires, the closer it comes to validating real motion.

Recall the RO vision on page 18:

> ... every `RealObject` *must have a location and speed* relative to the external coordinate system at any point in time.

This statement implies that the locations can be requested randomly and are not fired from the GPS itself. In order to replay a GPS trace, the motion between two points has to be filled in by the `GPSunit`. To do this, two new members are introduced and initialized with the unit

```
private double direction = 0, speed = 0;
```

Then, the `GPSunit` *needs* to register any *change* of constant motion. Physically speaking, a change of direction is an acceleration. With the code

```
speed = constant  and  direction = constant
```

the object is moving in a straight line and the current position can be determined by

```
distance = speed x (t1 - t0)
```

This means that the RO, owning the unit *has to* report every change of speed or direction to the `GPSunit` to get a correct trace. As long as speed and direction are constant, the `GPSunit` simulates the motion and is able to report the current `Position` at *any time*. This is where simulation and real-world operation completely differ—a GPS device does not move things:

```
GPSunit.move( speed, direction )
```

The RO simply says "From now on I am moving with *speed x* in *direction y*," while the GPS unit

- keeps track of the constant motion and can provide the `current GPSinfo` at any time;

- always moves from its `currentGPSinfo`;

- *does not* accept a `distance` or `destination` to stop moving;

- *only* accepts a new starting point by resetting the unit.

The implementation of `simulateMove` stores the last position and time provided externally and `getGPSinfo` is able to determine the exact location at any time thereafter. The "steering" is done by repeatedly using the public method `move( speed, direction )`. Real motion is unlikely constant, but using the average speed, we end up at the same place. The more points an object supplies the more realistic the motion will be. The implementation relies heavily on the `Position` implementation of `move`, `distance`, and `direction`, which implicitly validate each other.

4.11   Recording a `liveTrace`

To get the most out of this book and to learn how to program your own ROApp, the reader should consider buying or borrowing a GPS device and using it to scan his neighborhood, his way to and from work, a hiking trip—any activity of everyday life. By the end of this part of the book, you will get the chance to replay every trace you have collected.

Many GPS devices have some kind of internal memory and begin recording location and time as soon as they are switched on. The user has three modes to record traces: time, distance intervals, or a mixture of the two (auto mode). The automatic choice is advised, because the device will adapt to the speed. By recording a position every x seconds, the trace nicely shows the change of speed with different distances between points. Setting the unit to record every x meters can save a lot of memory as long as the device is not moving.

The life of the `GPSunit` begins with the instantiation, when the `new` operator calls the constructor of the class. The `GPSunit` should always record a `GPStrace`, since it can be useful in many ways. In the case of a car accident, it would be very helpful, if all involved vehicles had a precise GPS recorder (black box) on board. The traces could even reveal which car has violated traffic rules and caused the accident! A third party (RO) application would be able to replay the collision in slow motion and analyze it from different perspectives.

The implementation of a `liveTrace` is pretty straightforward. The trace is created with the `GPSunit` and the `initialGPSinfo` is added as the first element. In contrast to a real GPS device, the `GPSunit` is always aware of a change in motion, which makes the auto mode the easiest implementation. By adding `GPSinfo` with every change of motion, the trace has optimal compression, since points on a straight line don't add information. However, there is no point in explicitly adding a live trace thread.

In the context of an external observer a `GPSinfo` is added to the live trace whenever an external client requests information via a get request, i.e., `getGPSinfo()`. This should be helpful, when validating internally recorded data against the externally observed trace—and it forces the `GPSunit` to mark some points, even if it is moving at constant acceleration.

4.12   Loading a GPX File to a `GPStrace`

GPS traces can have many different sources (even simulations) and different formats. Earlier, the NMEA format was introduced with more generic GPS information than we need. The GPX format introduced in this section is much more suitable for our purposes and can usually be retrieved directly from a GPS unit.

Before playing back a GPS trace, it has to be loaded into the unit. In the development of the code so far, small traces were typed directly into the `main` methods.However, in order to access real traces stored on the hard disk, we need to deal with persistence and I/O.

> Please execute the method `NMEAconverter.main`; make sure to adapt the resource path to your environment. Note that some of the existing files will be overwritten.

The JavaGPS project comes with the small NMEA file `demo.gpslog` which can also be found on the book's web resources. The `NMEAconverter` was written to read and parse NMEA files, eliminate all sentences except GPRMC and GPGGA into `HDcastle.nmea` and write a simple text file holding the relevant GPS information: `HDcastle.csv`.

demo.gpslog

Since the NMEA converter resides in the default folder, the compiled class can be placed in any folder and can convert NMEA files via command line. Note the exception in the main method:

```
public static void main(String[] args) throws IOException
{
    String path = "D:/virtex/workspace/resources/gps/HD/",
        nameIn = path + "demo.gpslog",
      smallNMEA = path + "HDcastle.nmea";
    reduceNMEA( nameIn, smallNMEA );
      parseNMEA( smallNMEA, path + "HDcastle.csv");
      parseNMEA( smallNMEA, path + "HDcastleBody.gpx");
}
```

The NMEA converter creates a `HDcastleBody.gpx` file. This file (and later the map image of JavaGPS) will serve as a reference to compare the results with JavaGPS.

XML and Java in a Nutshell

The GPS eXchange format GPX is a widely available XML format supported by most GPS devices and applications. You can find traces at `www.openstreetmap.org` for working on development in your location. A `GPSinfo` element is represented as

```
<trkpt lat="49.40940666666667" lon="8.713693333333334">
    <ele>201.7</ele>
    <time>2002-03-01T13:40:30Z</time>
</trkpt>
```

The great advantage of XML is the strict definition of structures. XML documents are organized as a hierarchy of elements with start and end tags and were designed for ease of parsing. There are many XML parsers on the market. The SAX and DOM parsers are standards available in several different languages. In Java, you can instantiate the parsers by using the Java API for XML Parsing (JAXP) and parse XML documents through the Simple API for XML (SAX) and the Document Object Model (DOM) interfaces.

SAX is an event-driven model to parse serial file streams with good performance and low memory consumption. In the case of a connected GPS device, SAX can be used to fire events with every new reported position, but it is not designed to modify the data. In Chapter 7, *osmosis* will be introduced. Osmosis is built around a SAX parser, and you can easily install and study the sources in your IDE.

The DOM model is easier to use. It loads the entire GPX file into memory, and the application can randomly navigate through its structure, adding, removing, or modifying elements (and writing them back to the file system). The application can basically reorganize the data in a useful way. When the DOM parser reads an XML file, it builds a tree structure in memory and makes it available to APIs of the programming language. At the core of the DOM API are the `Document` and `Node` interfaces.

A `Node` is a general type that can hold different pieces of content. The
`org.w3c.dom.Document` class represents an XML document and is made up
of DOM nodes.

The basic functionality can be found in[8]

```
public boolean loadGPXfile( File gpxFile )
{
    Document traceDOM = loadGPX2DOM( gpxFile );
    :
    parseDOM( traceDOM )
    :
      extractGPXtraces( trkNodes );
      :
}
```

4.13   Play Back a `GPStrace`

After a GPX file is loaded into a `GPStrace`, the `GPSunit` replays the trace
with the first `GPSinfo` marking the beginning of the playback. The GPS
clock will *not* be set to the time stamps of the GPX trace implicitly. The
client of the `GPSunit` needs to set the clock to the environmental time explicitly (and usually only once in a lifetime).

The private method `simulateMove` developed earlier can be used to start
a motion with constant speed and constant direction. The simulation can
supply a `getGPSinfo` at any time. Replaying a GPS trace is simply a repeated call of this method—with precise timing.

To replay a trace as precisely as possible, a `ReplayThread` is created.
To replay a trace means looping through its elements and simulating the
move from one element to the next. The replay is started with the public
method[9]

```
public void replayGPStrace( GPStrace trace )
{
    :
    new Thread(new ReplayThread(), mode.toString()).start();
}
```

which calculates speed and direction, executes a `simulateMove` and stops
the motion (`speed = 0`) at the end of the trace.

4.14   The GPS Package: `roaf.gps`

The GPS package now has classes and interfaces to deal with geographical
coordinates and universal time (see Figure 4.1). Before progressing to

---

[8]See gpsUnitDemo(): gpsDevice.loadGPXfile(new File(nameIn)).
[9]See gpsUnitDemo(): gpsDevice.replayGPStrace(gpsTrace).

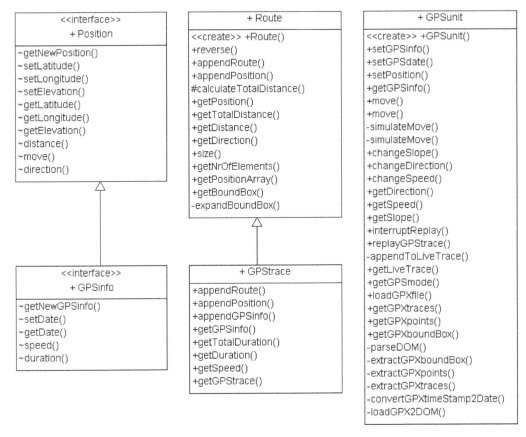

Figure 4.1. The GPS package to handle geographical coordinates, calculate metric coordinates, and keep universal time.

the next requirement, the `GPSapplication` was added to the package as a reminder of how to make use of its classes. Actually the `GPSunit` is a good entry point, and a `main` method could be added there. Adding a class provides an external view to the package.

4.14.1 Minimum Implementation and TODO List

When the architect constructs an application skeleton from scratch, he focuses on the final product, while programmers tend to drill down into details as they occur.

The GPS package was useful to introduce space and time and get an application running. Nevertheless, a lot of ideas pop up in the process, and, with the vision in mind, they should be saved for the actual roll-out of the application.

Additional features and TODOs. We list some of the additional ideas one might want to add to the bare-bones implementation.

- The GPS unit is a virtual GPS device. What if we want to replace it with a real GPS device (on a smart phone)? In order to do this, a `GPS` interface should be extracted from the `GPSunit`. Note that the interface should *not* include methods to set time and place, since this is done by the device itself.

- Replaying a GPS trace is useful. Yet looking at the overall vision, there are many more use cases for coordinating objects. Every (simulated) object should have ways to start, pause, and stop a trace. For example, in the case of a car accident, it is useful to be able to replay a complete scenario. A ROApp could record the motion of all participating ROs and use this data to analyze the scenario. Fast/slow, forward/backward, and pause methods could be used to implement the type of buttons found, for example, on any DVD player.

- For methods to be used internally by an RO, a proximity `Trigger` would be helpful. With this trigger, the journey of a car can be automated until it comes to an area where auto navigation reaches its limits and the car has to be driven interactively.

- A programmer could implement an XML validation and analysis of GPS traces: Do the timestamps make sense? Are they aligned in a sequential positive direction?

- A live trace could become really large over a unit's lifetime. How long does a lifetime last? For plausibility reasons, a ROApp might record all of its ROs' motions, while they are participating in a scenario. As in many online community games, a player can logoff from a scenario, save its current status, and shutdown.

  Therefor a trace management should be implemented to allow each RO to flush its trace to the file system and reload it at a later time. The RO user (or operator) should be able to specify a time span to regularly store the current trace and, vice-versa, to put it back together.

  Externally the client communicating with the RO should be able to say "Give me your trace between two time stamps, t_1 and t_2." Then, the ROApp server could compare the trace to the one recorded for the RO and check them against the scenario rules.

Chapter 5

From Geography to Cartography

5.1 Introduction

The `roaf.gps` package was created to manage spatial coordinates and to keep universal time. The `GPSunit` can be attached to any real object and it reflects its motion. From a developer's point of view, it is very helpful to wrap the GPS number-crunching into a single component. On the other hand, human perception is dominated by visual information, and it is more intuitive to display traces on a map rather than looking at large amounts of decimal numbers.

Cartography is the science of creating two-dimensional plane maps. This process can be realized with many different forms of *projections*. Each projection is a compromise between a two-dimensional plane and the actual three-dimensional surface one has to accept certain distortions.

Besides creating a (minimum) *mapping device*, the second objective of this chapter is to look at the mechanics of a framework. We will see how the `JComponent` is related to the Java *Swing Framework*, and how it can be extended to become part of the existing framework pattern.

5.2 Map Projection Systems

Geodesy, the science of measuring the Earth, makes use of geometric ellipsoids to idealize the actual geoid shape of the Earth, to apply clean mathematical equations and provide a latitude and longitude to identify any point on a given ellipsoid. (Trying to paste a rectangular sticker on a ball can illustrate the problems of mapping.) The ellipsoid is a three-dimensional body, which can not simply be folded to two dimensions.

Have another look at the JavaGPS application (Section 4.8) to find a list of geodetic Earth ellipsoids and Earth dates in the `COORD` class (including the Positions default ellipsoid-date pair for WGS84). All of these coordinates can be used to identify a point on the globe with (`lat`, `lon`) tuples—if the corresponding ellipsoid-date model is known.

For any chosen model, JavaGPS can transform the coordinates to the Gauss Krüger (GK) coordinate system. GK was developed at the beginning of the twentieth century and optimized to map Germany and surrounding countries. The global standard today, however, is the Universal Transverse Mercator system (UTM) which is similar to GK, except that the central meridians of the GaussKrüger zones are only 3° apart, as opposed to 6° in UTM. As a consequence, the scale variation within a GaussKrüger zone is about 1/4 of what it is in a UTM zone.

For project-oriented development, it makes sense to look at the bounding rectangle of the example provided in JavaGPS. The map image `HD.jpg` shows the famous castle in Heidelberg and can be described with two sets of coordinates:

The constructor `org.iu.gps.MapPanel()` uses cartographic GK coordinates:

```
bitmapRect.x = 3477970; bitmapRect.width  = 2029;
bitmapRect.y = 5475590; bitmapRect.height = 1939;
```

These GK coordinates can be translated to geographic positions with

```
   COORD.convertToLL( 3477970, 5475590 );
-> 49.41791753327985, 8.69634211847276  - upper left
   COORD.convertToLL( 3477970 - 2029, 5475590 - 1939 );
-> 49.40040474786639, 8.668492617517886 - lower right
```

Now `GeoPoint.distance` can be use to determine metric coordinates. The horizontal scale is from 0 to 2,022 meters and the vertical ranges from 0 to 1,947 meters. So the area of the map image roughly covers 2×2 kilometers.

5.3   Requirements for a Map GUI

A graphical user interface (GUI) can be used to visualize information for a better and more intuitive understanding. The ROAF, however, will be designed to operate without a graphical representation. Nevertheless, the data structure should enable a mapping front end to be attached and display the graphical representation of the application. To document that the GUI is not an essential part of the `roaf`, the GUI package name will be labeled with an `x`, i.e., `roafx`, similar to the usage in the Swing Framework in `javax` packages.

The three coordinate sets, GK, (`lat`, `lon`), and (x, y), can be abstracted to a simplified mapping device. Each coordinate set are decimal numbers describing a horizontal and a vertical scale of a Cartesian plane. The directions west and south are indicated by negative values.

The common expectation of a map's proportions is to match a picture of a city taken from a helicopter or a satellite. The x- and y-scales are identical regardless of the picture's orientation. A football field should always be a perfect rectangle, and a circle should not be distorted. A car moving with a constant speed should be displayed at constant speed on the map of a navigation system. Nevertheless, the horizontal and vertical scales have to be independent of each other, since latitude and longitude scales are not identical in terms of distances between two values (see page 36). As a starting point, the *map image* `HD.jpg` will be used as a valid map with a correct projection of the geographic area.

The mapping component should be able to

1. display the area given by a horizontal and a vertical set of decimal values;

2. draw a coordinate grid based on the four decimal values;

3. fit a map image, also described by two sets of decimal values, into the mapping area (the grid);

4. draw various geometric primitives to represent points (locations), lines (road segments), polylines (routes and traces), and polygons (areas).

The geometric primitives indicate formats of digital maps, which are not stored as images, but rather as chunks of information. Digital maps usually do not supply cartographic coordinates. Instead, they are built with geographical (`lat`, `lon`) coordinates usually based on the WGS84 ellipsoid. Processing digital map data will be introduced in detail in Chapters 7 and 8. For the time being the map image will do.

5.4   The Framework Pattern

The object-oriented paradigm has produced a lot of standard approaches to common programming problems referred to as patterns. Although there is no exclusive "framework pattern," the Swing Framework can help to supply a deeper understanding of the envisioned ROApp Framework.

The Java Swing Framework is a general-purpose graphical user interface. The framework launches its own drawing thread to draw windows and their subcomponents. A customized component is created by extending the `JComponent` to become part of the existing framework by inheritance. The framework propagates external events from the user or operating system to all subcomponents. Every component can override Swing methods with dedicated code to trigger these events internally.

The object-oriented lesson to be learned from Swing is the potentially random access of one object from different contexts. The vision of a framework for real objects is analogous. A `RealObject` is envisioned as the platonic object to be placed in the framework. The fundamental object automatically takes care of the communication and the programmer implements behavior inside this development environment.

It is also important to understand that the framework and the basic component can be developed highly independently of one other. The Swing Framework provides a large number of components specializing the base component in a hierarchical manner. The idea of the `RealObject` is to provide a base class for any real-world object with pre-implemented physical attributes and automated positioning mechanisms.

> If you are not familiar with the Swing Framework, you should study the `roaf.book.map.gui.SwingApplication.main` method. This application demonstrates how to create `JComponent`s and how to fit a variable grid into a component.

The development of a map panel will be based on the `JPanel` extending the `JComponent`. In the long term, a ROAF developer choosing to create a traffic application might prefer to use a predefined `RealCar` extending the `RealObject` to focus on implementing a navigation system to steer the car. By connecting the `RealObject` to a traffic application, it can listen to the traffic to adjust the speed.

5.5 Creating a `MapPanel`

> The remainder of this chapter will walk you through the development of the `roafx.gui.map` package. In Section 5.8, a GPXviewer is rolled out to demonstrate the usage of the mapping components with the GPSunit. The GPXviewer resides in `roaf.book.map.gui`, and it is also deployed as an executable jar file in the `../resources/deployed` folder together with instructions in the `GPXviewer.txt` file.
>
> It is suggested to go through these instructional steps now.

The first step in creating a `roafx.gui.map.MapPanel` is to extend a Swing component and add the vital information to the constructor:

```
public class MapPanel extends JPanel {
  public MapPanel( Rectangle2D.Double gridArea )
    { this.gridArea = gridArea; }
    ⋮
```

The rectangle indicates the task to draw a Cartesian coordinate system on the provided screen space. The parameter `Rectangle2D` in the constructor defines the horizontal and vertical (`from`, `to`) pairs for the grid. The intuitive next step is to calculate and draw the grid on the screen. At this point, the developer has to be aware that a Swing component can be constructed at any time—yet, it is impossible to request its actual size before it is added to a top-level container. Only after all components have been supplied, Swing can start laying out the components with `frame.pack()`.

Without the grid defined in the constructor, there is no way to draw geometric primitives. Therefore the programmer has to listen to the communication inside the framework. The `setBounds(x,y,w,h)` method is invoked on the `JComponent` by the Swing Framework whenever the assigned space has changed. By calling the `super` method, the framework can proceed with its normal procedure, and the specialization can be added afterwards.

```
/* trigger the (re)calculation of mapscales.*/
public void setBounds(int x, int y, int width, int height)
{
  super.setBounds(x, y, width, height);
  // add customized code here
  calculateScales();
    :
```

Now, the `calculateScales()` method has access to the component's pixel size and the decimal grid, which is the prerequisite to draw with decimal values. The method will not be described in detail. Basically, the horizontal and vertical scales are being managed by additional visual Swing components using predefined Swing constants:

```
class MapScalePanel extends JPanel implements SwingConstants
```

The main purpose of the two map scales is to provide a conversion of x- and y-values from pixel to decimal and vice versa. The two methods return the (x, y) pixels for spatial (`lat`, `lon`) coordinates:

```
int horPixel = horScale.dec2pix( pos.getLongitude());
int verPixel = verScale.dec2pix( pos.getLatitude() );
```

Vice versa, any position on the screen can be calculated to real-world coordinates:

```
public void mouseClicked(MouseEvent e) {
  double lon = horScale.pix2dec(e.getX());
  double lat = verScale.pix2dec(e.getY());
    :
```

The methods take into account that Swing counts vertical pixels from top to bottom (reading direction), while the (longitude) values grow from bottom to top.

Note that the map scales are only visible from inside the package, since they have to and should only be used in conjunction with the map panel.

Creating the scales as visual components opens the option to place them inside a Swing Container and draw them.

With these conversion methods, any GPS position can be converted to spatial coordinates. The next "hook" (a way for customizing the handling of some event or process) to Swing is the method that actually paints the component:

```
protected void paintComponent(Graphics g)
```

+ MapViewer

-menuLoadImage()
+adjustBigMapRatio()
-menuLoadGPXbox()
-createMenuBar()
-createGPXloader()
+mapClickedAt()
+mouseWheelMoved()
-changeBigMapSize()
+mapPortChangedTo()
-createMapPanels()
-layoutComponents()
+<u>createAndShowGUI()</u>
+<u>main()</u>

<\<interface\>>
+ MapMouseListener

+mapClickedAt()

<\<interface\>>
+ MapPortListener

+mapPortChangedTo()

+ MapPanel

<\<create\>> +MapPanel()
+setGridArea()
+getGridArea()
+showGrid()
+drawImages()
+drawBoundBoxes()
+fillPositions()
+drawPositions()
+drawEdges()
+drawRoutes()
+setBounds()
+setPreferredSize()
+addMapMouseListener()
-determineGridRectangle()
-calculateScales()
-dec2pixPositions()
-dec2pixEdges()
-dec2pixFullPositions()
-dec2pixRoutes()
-dec2pixImages()
-dec2pixBoundBoxes()
-dec2pixRectangle()
~pix2decRectangle()
#paintComponent()
-paintImages()
-paintGrid()
-paintBoundBoxes()
-paintPositions()
-paintEdges()
-paintFullPositions()
-paintRoutes()

+ MapScrollPane

-<u>serialVersionUID</u>
~mapPortListener
-mapPanel

<\<create\>> +MapScrollPane()
+showMapScales()
-setCornersAndHeaders()
-cornerButton_actionPerformed()
+getMapPort()
+addMapPortListener()
+getMapCenter()
+centerMapTo()

~ MapScalePanel

<\<create\>> ~MapScalePanel()
-setOrientation()
~setPreferredLength()
~setScale()
-calculateScale()
~dec2pix()
~pix2dec()
#paintComponent()
-drawScale()
~<u>roundDouble()</u>
-getMagnitude()

Figure 5.1. An overview of the `roafx.gui.map` package to build a minimum ROAF GUI front end. The drawing of geographical coordinates and map primitives takes place in the `MapPanel`. The transformation from decimal (geographical) coordinates to pixel values is calculated in the `MapScalePanels`.

Note that this method can only be overridden inside the inherited class due to the `protected` visibility. Again the method is invoked by the framework, when it is the component's turn to paint itself. The `Graphics` class provides methods for the remaining requirements to draw a (map) image and geometric figures:

```
drawText()      drawString()
drawPolyLine()  drawLine()
drawPolygon()   fillPolygon()
drawRect()      fillRect()
```

The implementation of these methods can be found in the API of the `MapPanel` (see UML diagram in Figure 5.1). The prefix `paint...` is used in the component's context to refer to `paintComponent` with a `Graphics` instance as the argument, while the prefix `draw...` refers to the `Graphics` drawing methods.

Now, the `MapPanel` can be used to display geometries on a map with GPS positions. Note that all drawing elements are passed to the map panel with fixed-size arrays, the simplest and fastest collections. This design enforces an external management of the arrays. To add or remove elements, a new array has to be passed again to `draw` methods to refresh the map display.

5.6   Creating a Swing Mapping Application

The advantage of creating a Swing `MapPanel` is that every programmer familiar with Swing can create his personal Swing application and add a map in the same way as adding any other component. For example, the map can be placed in the Swing containers `JScrollPane` and `JSplitPane`.

Of course, Swing has its own specifications; for example, a beginner might be confused when trying to position his components explicitly. Generally speaking, layout managers (over)rule size and location of Swing components placed inside containers. Although Swing allows you to turn them off, it is not a good idea, since you loose a lot of functionality required for windowing systems. This is especially undesirable when the GUI needs to fit the screen size of hand-held devices.

When displaying a map (image), resizing of the map can be irritating to the user. It also consumes CPU power by rescaling every map item. We therefore add another Swing component to the map package: the `MapScrollPane`. It extends the `JScrollPane`.

THE JAVA TUTORIAL > CREATING A GUI WITH JFC/SWING > SCROLL PANES

A JScrollPane provides a scrollable view of a component.
When screen real estate is limited, use a scroll pane to dis-

play a component that is large or one whose size can change
dynamically.

With the `JScrollPane`, the `MapPanel`s size can be adjusted to the original image size and will not be changed by layout managers. The map
component's preferred size is the fixed size and remains fixed as long as
it is not changed explicitly. The viewing experience is more comfortable,
since the user can adapt his perception to the map details more easily—
regardless of window and screen size. Thus, screen and map panel can be
managed independently. The `MapScrollPane` is basically a higher-level map
component implicitly managing the `MapPanel` and its `MapScalePanel`s. Since
the scrollable component *has to have* a preferred size, the Swing method
`setPreferredSize` is overridden to invoke the `setBounds` method, the (one-
time) trigger to draw a map.

5.7   Interacting via `MapEvents`

When developing libraries, some logical conflicts might occur. For example,
external clients of Swing might need information of the things going on
inside the framework. We need to find a way for a Swing developer to
provide information to a client not existing at development time?

One way to do this was already demonstrated by inheriting a Swing
component (`MapPanel extends JPanel`), overriding a method (`setBounds`),
and using the method body to trigger any other process (`calculateScales`).

Another way to accomplish the same functionality is to use `Events` and
`Listeners` as described in the Java Tutorial:

THE JAVA TUTORIAL > CREATING A GUI WITH JFC/SWING > EVENT LISTENERS

> Any number of event listener objects can listen for all kinds
> of events from any number of event source objects. For exam-
> ple, a program might create one listener per event source. Or
> a program might have a single listener for all events from all
> sources. A program can even have more than one listener for
> a single kind of event from a single event source.

By default, the `MapPanel` can be scrolled inside the `MapScrollPane`. For
convenience, it would be a nice feature to be able to drag the map with the
mouse. More precisely, one might press and hold the mouse button and
move the mouse.

In order to retrieve these events from Swing, the scroll pane could be tagged as `MouseListener` and `MouseMotionListener` and implement the methods needed to move the map:

1. mousePressed > 2. mouseDragged > 3. mouseReleased

The actual implementation uses the `MouseInputAdapter` to avoid empty method bodies. The dragging functionality is achieved with the help of event triggers, completely hidden inside the scroll pane. The user should not be bothered with the conversion of pixel to decimal values. Instead of returning the cursor position in pixels, it should return the decimal position.

The idea is to implement a mouse-click listener `MousePosition` to catch the mouse-click event `mouseClicked(MouseEvent e)` from Swing. At this point, the `MouseEvent` can be extended to a `MapEvent` to add the decimal coordinates. To restrict the number of classes, we create a `MapMouseListener` with only one method:

```
MapMouseListener extends EventListener
{
    void mapClickedAt( MouseEvent e, Point2D decPoint );
}
```

From the developer's perspective, the arguments of a listener method are actually the return values from the event. The implementing client can use the information about the mouse event provided by Swing *and* the decimal point clicked on the map. The pixel-to-decimal conversion is hidden in the `MapPanel` in order not to bother the map user with pixel coordinates.

A client, like the `GPXviewer`, using a `MapPanel`, only has to execute these steps:

```
// 1. declare class as listener:
public class GPXviewer implements MapMouseListener
{
    mapDisplay = new MapPanel( gridArea );
    ...
// 2. direct mouse events to this listener class:
    mapDisplay.addMapMouseListener( this );
    ...
// 3. 'catch' decimal coordinates of mouse click:
    public void mapClickedAt(Point2D decPoint)
    {
        String lat = decPoint.getY() + "";
        String lon = decPoint.getX() + "";
        ...
//     propagate coordinates to listener
    }
}
```

Another use of listeners is to report changes. A client can get the decimal view port any time by using `mapScrollPane.getMapPort`, but also needs to be informed when the port changes if the user interactively (randomly) scrolls or drags the port. The following steps need to be executed:

```
// 1. create listener
public interface MapPortListener extends EventListener {..}

// 2. add viewport change listener
public synchronized void
      addMapPortListener( MapPortListener listener ) {..}

// 3. catch change event from Swing and propagate to listener
private class MapPort implements ChangeListener {..}
```

The components `MapPanel`, `MapScalePanel`, and `MapScrollPane` are specialized Swing components made to assist developers in visualizing positions, routes, and bounding boxes in geographical coordinates. By applying inheritance, the code used by the Swing framework can be accessed as intended.

Again, the developer should always keep the two perspectives in mind: the object's internal coding and its external behavior. A Swing component perfectly describes this situation. The component's internal calculation is visualized by the grid, and externally the component's size can be modified by the user (mouse), by Swing (`setBounds`), by the Swing application (`setBorder, setBackground`), or even by the operating system (arrange windows), simultaneously. Casting components for each environment supports a clean design. By casting the `MapComponent` to `JComponent` Swing can only apply the methods pre-implemented in the framework.

5.8   Deploying the `GPXviewer.jar`

The GPX viewer is a useful application to visualize GPS information retrieved from GPX files. The `GPXviewer.java` file combines the components `MapPanel` and `GPSunit`. The reader should go through the code to identify the GUI features and how the components are integrated.

To support developers in visualizing GPS information, the GPX viewer can be deployed to a stand-alone application:

THE JAVA TUTORIAL > DEPLOYMENT > PACKAGING PROGRAMS IN JAR FILES

The Java Archive (JAR) file format enables you to bundle multiple files into a single archive file ...   how to customize manifest files to set an application's entry point.

Find the `../deployed/GPXviewer.jar` and launch it with the command line:

```
java -jar GPXviewer.jar
```

On many Java installations, you can launch jar files with a double click. If you are new at this, it is not advised, since the viewer does not have a status bar. Therefore, you should keep an eye on the command line to receive feedback and errors.

Run the quick start instructions in `GPXviewer.txt` in the same folder as the viewer. It should be easy to see that the application is not ready to be marketed. If there is there something you think is missing or something you don't like, you should launch the viewer in your IDE and add it and/or change it. There is room for improvement.

Chapter 7 (page 79) introduces the open source map viewer JOSM provided with the OpenStreetMap project. However, you might want to search the web to find a different tool to create GPX files, and then organize a map image of your environment and clone the HD folder to it.

5.8.1   Minimum Implementation and TODO List

In the project context, the `MapPanel` is a minimum implementation for converting `dec2pix` and `pix2dec` inside the map scales and for drawing geographical information. From another point of view, the panel contains a number of arrays, which can be filled from an application. Although there are many (free) tools to handle GPX data, they are not suited to draw Java arrays or replay traces.

Here is a sample TODO list. You might want to add some additional items.

- Extend the GPS unit to read, modify, and write GPX files to disk.

- Beautify graphics with

> **`java.awt.Graphics2D`**
>
> This Graphics2D class extends the Graphics class to provide more sophisticated control over geometry, coordinate transformations, color management, and text layout. This is the fundamental class for rendering 2-dimensional shapes, text and images on the Java(tm) platform.

5.9   The GUI Mapping Package

The UML diagram in Figure 5.1 gives an overview of the `roafx.gui.map` package. The reader should experiment with the `roafx.swingmap` package to avoid conflicts with the (soon-to-be-developed) software built on top of the displayed package.

Part III

ROs - RealObjectS

Parts I and II provided the prerequisites to grasp the project vision as well as a feeling for the object-oriented programming approach. The GPS unit and the map panel encapsulate domain knowledge and provide developer tools as a common basis.

This part implements `RealObject`s as the root class for all *real-world objects* interacting in *real-world scenarios*. In the real world, a physical object is any thing with mass, size (shape), and a location (orientation) at any point in time. In the computer world, every (real) object is a program of its own, an entity with an internal *behavior* and interfaces connecting it to *external environments*.

In order to reflect realistic motion, the platonic `RealObject` will be enabled to replay a GPS trace recorded in the real world. Before a `RealObject` can simulate real motion it has to be enabled to read a digital map and move along its network of roads.

The processing of digital maps for cartography and navigation is described in detail. We make use of the freely available digital map source OpenStreetMap (`www.openstreetmap.net`). With preprocessed digital map data for a dedicated area, each `RealObject` can be equipped with a navigation system to read the map and make a decision on which direction to take.

Chapter 6

Objects in Motion

6.1 Introduction

After a systematic analysis of the vision and some preliminary coding to support ROAF developers, we tackle the implementation.

Although the GPX viewer can replay a trace on a map, the observer has no clue *what* is actually moving on the map. The GPS device will be tied to a real object as a skeleton to implement behavior and provide additional attributes to the external world. The most straightforward implementation of a realistic object is to internally replay a GPS trace recorded in the real world.

> The implementations described in this chapter can be found in the `roaf.book.ro` package.

6.2 Every `RealObject` has a `GPSunit`

The book's vision was derived from the `java.lang.Object` as the origin of every new class to be built on top of it and the `RealObject` as the basis build any physical elements. According to the initial RO vision (page 18), any physical (real) object has to have a location, a mass, and a body; vice versa, no thing is real without these attributes. The software designer can enforce these attributes at creation time with an adequate constructor:

```
RealObject( GPSinfo gpsInfo, double mass, Shape shape )
```

To supply a location each `RealObject` has a hidden virtual `GPSunit` to keep track of GPS coordinates. The `GPSunit` is a private member and cannot be accessed externally—by design. The RO is responsible for its own (auto) motion: The decision to replay a recorded trace, to simulate motion or to hook up a real GPS device (or a mixture) is private to the RO, and the location can not be accessed and modified externally. The idea of the ROAF Turing Test is to obscure these different modes to external observers.

To get a grip on the relationship (aggregation) of the `RealObject` to its `GPSunit`, a designer can describe a *use case*:

1. Imagine you (a body with *location*, *mass*, and *size*) represent the `RealObject` carrying a cell phone (a *reference* for other people to reach you and inquire information) and a `GPSunit` (*hidden* in your jacket).

2. You are standing on or near a bank of a river (location) and you turn on the GPS device, which is attached to you and retrieves *your* geographical `Position` and the universal time (`GPSinfo`).

3. A friend calls you and ask, "Where are you?" You answer, "I'm at Regen street parallel to the Regen river" (local knowledge). To be more precise, you look at the GPS and say, "I'm at `N49.031, E12.103`." He asks, "What time is it?" You read it on the GPS, the most precise clock available and reply, "Let's meet downtown at the cathedral."

4. You begin to walk downtown to meet your friend. The `GPSunit` refreshes your position approximately every second and logs every `GPSinfo(lat, lon, elev, time)` to its internal memory.

5. You're late and jump into a cab. The GPS is now logging the cab's trace along the streets.

The design process can simply follow this *use case*:

The constructor

```
RealObject( GPSinfo gpsInfo )
{
    gpsUnit = new GPSunit( gpsInfo );
}
```

sets place and time (of birth). The `Position` of the `RealObject` can not be set externally—you can not move a body by supplying coordinates like in Star Trek. Similarly, people usually guide each other by names of streets or places. The `GPSunit` records the RO coordinates, while the RO can describe its motion in terms of phrases like "turn left at the next intersection (where you can see the cathedral), and then walk straight in the direction of the cathedral."

Whenever exact position or time are needed, read it from the `GPSunit`:

```
final public GPSinfo getGPSinfo() // where am I and what's the time?
{
    return gpsUnit.getGPSinfo();   // ask my internal GPS
}
```

Some methods are marked as `final`, the object-oriented technique for avoiding overriding and also to ensure that all ROs are really using their internal `GPSunit`.

The method

```
final public double getOrientation()
{
    return gpsUnit.getDirection();
}
```

demonstrates how terminology can be adopted for different objects and contexts.

Speed (velocity) is intentionally *not* provided externally, since it depends on the RO's private `Position` implementation. When you get caught speeding with your car, you can't argue that *your* speedometer was in the legal range!

At this point the RO can become any thing, not only a moving thing. Of course, a building has no speed relative to the ground (`Position = const`) and logging GPS tracks wouldn't make much sense. Yet, the building does have a location at every point in time, which will be provided by the GPS unit and can be synchronized with or by the framework.

6.2.1 Every `RealObject` has a `Shape`

The `Shape` provided in the constructor on page 69 refers to the body of the object. During the project setup, this shape should be kept as simple as possible. For example, it can be constructed with

```
public Shape( double length, double width, double height )
```

For the time being, the three dimensions should describe the outer box of the object. This is evident for buildings. For busses, the orientation should be considered in a two- or three-dimensional simulation. For motorcycles, the outer shape would need dynamic adjustment, since it gets lower and wider when leaning into a curve. A human body is even more complex: lying, sitting, standing, running, etc.

6.3 abstract Motion of a `RealObject`

The RO is created in conjunction with its GPS device to provide a heartbeat. Then the real object is ready to move (or to be moved). Physically speaking, any motion is described by the ratio of distance and duration in a direction. People move by taking one step at a time and accelerate by increasing the step's lengths and frequency. Motor vehicles are driven by engines transforming fuel (stored energy) into motion and are directed with the steering wheel. Every object has a different way to move and makes a concrete implementation of motion impossible. The only constraint so far, is that the `GPSunit` has to reflect every change of motion with

```
final protected void move( double direction, double speed )
{
   gpsUnit.move(direction, speed);
}
```

Although this `move` is moving the `gpsUnit`, the RO programmer has to carefully distinguish real and simulated motion. The method is `final` to force every object to report *the result* (or return value) of a simulated motion relative to the external environment with `.getGPSinfo()`. Consequently, this method is used to write (and log) the result of a simulation (or playback, or live trace) step and the `RealObject` can be moved with

```
... protected void move()
{
//    1. simulation calculates next move  ...
// or 2. retrieve move from trace replay  ...
// or 3. retrieve move from a real object ...
//    => finally execute move with values
      move(direction, speed);
         :
   }
}
```

Thus, the move will be logged to the RO history and, externally, the coordinates can be retrieved at any point in time.

By repeatedly calling the move method, the object can be guided along any path in space and time: a person walking, a jet plane flying or a fish swimming. This usage of `move` adheres to the mathematical description of motion:

$$\Delta \, \text{distance}/\Delta \, \text{time} = \delta s/\delta t.$$

Since the ROAF is meant to be a real-world simulator for *every* existing object, it is impossible, at this point, to implement a concrete way to move. In object-oriented programming, this problem can be overcome by adding an `abstract` method:

```
abstract protected void move() { /* no implementation! */ }
```

Abstract methods do not have an implementation and as soon as a class has an abstract method the whole class has to be declared `abstract`:

```
public abstract class RealObject
```

In general, Java allows three kinds of classes: full implementation, partial implementation, and no implementation. The `Position` interface is a class without implementation, `GeoPoint` is a class with full implementation.

The `RealObject` is a class with a partial implementation and can not be instantiated with the `new` operator. The `protected` visibility of the method move *forces* every RO programmer to extend the `RealObject` in order to use it. Real-world semantics are modeled with the mechanisms of visibility and class type.

6.3.1 Every RealObject Has Mass and Inertia

The abstract method `.move` provides a programming shell to supply motion parameters from the internal behavior. On the outside, observers can sample traces to validate them against the environment (map) and check plausibility. In terms of the vision, RO and ROApp programmers see the real object from different perspectives.

Handling mass and inertia was mentioned earlier (see page 15) and is discussed in more detail on the website (www.roaf.de) under the heading "Objects and Physics." A car, just like any mass object, can not simply set the speed. The `move` method has to be improved in order to accelerate and decelerate according to the object's power and mass. Inertia holds the mass back, while the force pulls it forward. Since the mass is part of the `RealObject`, the change of motion can be calculated roughly.

With a set of real traces for a specified vehicle, the acceleration can be reverse-engineered for different speed ranges. The smallest curve radius in relationship to speed can be determined. For realistic behavior, according to the rules of force and inertia, the RO programmer could write a trace analyzer to identify extreme situations to define the characteristics of the vehicle and the roads. The simulation could use these vehicle rules for a realistic drive—on the analyzed roads.

Physically speaking, the observed motion needs to take into account mass, inertia, and external parameters like surface, weather, etc. For the implementation of a `RaceCar`, a programmer might implement the motion through certain formulas when a user races a car.

Since the external observer (a server application) can only validate *how* your object moves without checking the code, the RO programmer can choose any implementation with which he is comfortable.

Here is a little game you can play to get an understanding of inertia and change of motion (speed and/or direction) without applying the actual physical formulas. First, you need a piece of graph paper which will serve as our Cartesian coordinate system. Draw a racing circuit by hand with a space between the curbs from three to five squares and a start/finish line with two black dots representing two race cars.

To start the game, each player can move his car (black dot) one square forward or along the diagonal to the left or right. The black dots are connected with vectors (arrows) as in Figure 6.1. For the next move, copy

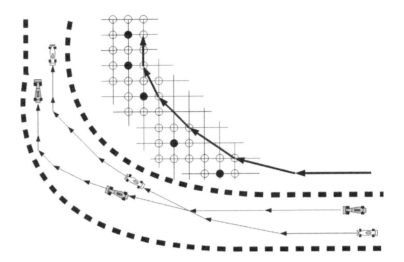

Figure 6.1. Motion can be calculated with geometric rules instead of physical formulas.

the previous vector in the same direction and same length and draw a small black circle. The corners of the four squares around this circle define nine possible destination points (black circles). Pick one of these in the appropriate direction to match your race track and draw the next vector for your next move.

For a simple game, nine possible destination points will be adequate. For a real-world simulation, the destination square is a characteristic ellipse changing its shape according to the car's parameters. If the car is going full speed only the back part of the ellipse represent the destination area and the ellipse is very thin. The car can only make slight turns at high speeds. When the car reduces speed the ellipse gets wider.

The upper part of the figure (thick arrows) demonstrates the rules for moving your car. The thick dashed lines show how you can actually analyze typical racing maneuvers.

Let's say both cars have the same top speed entering from the right side. The gray car is keeping up its speed to defend the inside of the next curve. The white car realizes that it can't pass the opponent and reduces speed.

Due to mass and speed (inertia), the gray car is carried away from its ideal course. Due to a lower speed, the white car can pass its opponent. Note that the gray car can get the lead back, if the next turn is sharp to the right.

This is only one way to provide the two parameters *speed* and *direction* needed for the move method; you should try to implement your own. As a starting point, you could analyze your recorded traces with your vehicle on the way to work.

In conclusion, mass and inertia need no extra methods for a `RealObject`; the `Shape` should be taken into account in the racing scenario. Overtaking can only succeed without the cars touching one another. The implementation of the `MassObjects` on page 13 showed that two mass objects have to "know" each other's size in order to determine the point of the collision. This should be implemented with the Java `Listener` concept. Each `RaceCar` would be required to implement listeners, or in real-world terminology, sensors. The collision `Event` could provide the momentum (speed × mass) and the point of impact. As in the Swing framework, events can communicate independently, internally, and asynchronously, bypassing the application flow.

6.4   Creating a Motorcycle

An advantage of having your own development environment is the privacy of coding step-by-step. Thus, the constructor

```
RealObject( GPSinfo gpsInfo ) // provide space & time
```

will be sufficient at this stage to continue development based on traces recorded in the real world. The ROAF Turing Test (see page 18) strongly supports `roaf` development and validation: recording a trace in the real world is the perfect top-down guideline for a bottom-up development of a realistic simulation. Since authentic GPS traces represent scientific facts, they are perfectly suited as a guideline for the simulation mode. Any RO (vehicle) has to be able to follow a GPS trace (internal). The idea is to technically decouple RO and ROApp development and yet synchronize all ROAF components in *one* reality. Each developer should have a high degree of freedom to build his own applications—with reality as the only constraint.

The traces in `../resources/gps/RGB-BUELL-NW` were recorded for one (instance of a) motorcycle and reflect the bike's characteristics.[1] Most rides were purely joy rides: no rush, small and clean roads, plenty of turns, and sunny warm weather. All traces are authentic and will guide us in the next steps.

Our very first `RealObject` will be a `Buell_XB12Ss`:

```
public class Buell_XB12Ss extends RealObject
{
    public Buell_XB12Ss( GPSinfo gpsInfo )
        { super( gpsInfo ); }
    protected void move() { }
}
```

---

[1] The specification of the bike can be found at **www.buell.com**.

Simply let the IDE override the constructor and add the `.move` method. That's it. Note that although the Java compiler forces you to supply a `.move()` implementation, it does not prevent the stub to be left empty.

Since a GPS trace was identified as a realistic real-world simulation, seen by an external observer, the motion is derived from the GPS unit's playback method described in Section 4.13. To make use of the playback, two methods are added to the `RealObject`:

```
final protected void replayGPStrace(     File gpxFile )
final protected void replayGPStrace( GPStrace gpsTrace )
```

Note that the methods are not added to `Buell_XB12Ss`, since they can be applied to any `RealObject`.

In order to move a vehicle you need to be in the driver's seat. For the moment, the `main` method will serve as the environment to create a Buell motorcycle. It's a plausible development environment as it can access `private` and `protected` methods, i.e., it can be controlled from inside.

6.5   Observing Motorcycles

`MapViewer`. As an application grows, the number of classes increases and it's a good practice "to close some doors behind you." In the prototyping

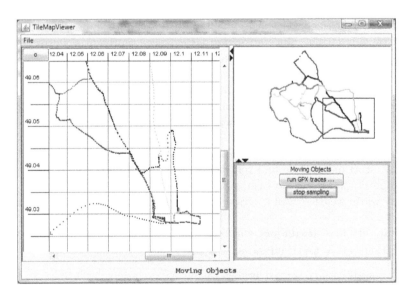

Figure 6.2. The `(Tile)MapViewer` displays a large map for zooming and a mini-map for an overview of the complete bounding box. The `MovingObjects` panel (lower right) can be used to replay GPS traces stored in GPX files.

process, the developer should omit some of the implementation detail and build a minimum implementation in order to build a higher-level application, but implemented in such a way that vital missing details can be added easily. In Part II, a mapping GUI was developed from inside the `roafx.gui.map` package. From here on the RO developer wants to focus on the non-visual development, while using a `MapViewer` to monitor the actual movements (see Figure 6.2). The `MapViewer` is a GUI similar to the `GPXviewer` with slightly different design intentions. By using the static method,

```
MapViewer gui = MapViewer.createAndShowGUI();
```

the viewer can be reduced to a member variable in another application.

`MovingObjects` is the development environment for the ROs and even the closing of the `MapViewer` should not affect other objects in the application. For isolated map analysis, the `MapViewer` can always be launched with its main method or as an executable `jar`.

In a team, the GUI developer can distribute a `MapViewer.jar` and the RO developer can basically work against the `MapPanel` API and apply listeners to talk to the environment, whenever the user interacts. The `controlPanel` can be retrieved to add application-specific controls (as we will soon see).

`MovingObjects.` Please go through steps *1* through *3* in the quick start instructions[2]

```
../deployed/MovingObjects.txt
```

`MovingObjects` is the first small real-world scenario, or real-object application (ROApp) and provides

1. a GUI (variable) to observe `RealObjects`;

2. an environment with time and coordinate implementation;

3. a button to load traces, create objects (motorcycles);

4. a button to start a sampling process of the object's current positions.

The `RealObject` is only at the beginning of its evolution and was marked **abstract** right from the start and *has to be* extended for further development. The `RealObject` was instantiated with the help of the `Buell_XB12Ss` class.

Technically speaking, the application collects the object's `Position` in an array to be visualized in `MapPanel.paintComponent`. While the `GPXviewer`

---

[2]Step *4* will be covered in Section 7.10, creating map tiles.

had few performance requirements, the new environment allows scaled performance tests. Observe the GUI behavior with 20, 30, and 40 traces, while constantly changing the window size, map position, etc.

The GUI developer can improve the map application by increasing the number of traces, reducing the drawing on the visible part of the map, add a preference pane as in the `GPXviewer`, etc.

Watching the bikes in the `RGB-BUELL-NW` scenario illustrates a number of things: The GPS receiver collects reliable positions on the same (side of the) road on different days. Not all, but most of the points line up to a plausible trace. The implementations of `Position` and `GPSpoint` allow a reconstruction and playback of the traces. The traces closely match the roads on the map image and, with constant sample rate, the pixel distances imply speed.

The map `RGB-BUELL-NW.png` of this scenario is sufficient information for anyone familiar with this area. Most of the traces are in the area bounded by three rivers, covered with plenty of woodland, grass or fields (green), downtown in the south is roughly depicted by a gray area and the train track gives the image a signature. Most people know the shape of their country or state (in a familiar projection) at first site. For another scenario, this map may be completely useless. Images of maps or raster maps, are composed of a fixed number of picture cells, pixels. These maps are totally inflexible and although they can be scaled, as in the `MapPanel`, the details do not change.

Where do these map images come from? The next chapter will guide you through the creation of your own map for your geographical area and purpose.

Chapter 7

Processing Digital Maps

7.1 Introduction

In the last chapter, traces were displayed on top of a map *image* for an observer to visually validate the trace. For a computer, these images are practically useless, and it would be almost impossible for the computer to identify roads. A machine can only interpret digital maps describing the mapped area by structured information —often referred to as *spatial data*.

In the context of the ROAF project, digital maps are vital for the orientation of real objects. Digital maps bind real objects and real-object applications together. Externally, the map describes the object's environment, while the object can internally *make a decision* in which direction to go.

This chapter describes how to process digital maps for dedicated applications with a *map compiler*. At the open source project `openstreetmap`, you can find free digital-map data for your geographical area to set up your own OSM map compiler.

7.2 Overview

Part II introduced the basics of geodesy, how to match geographical coordinates with the globe's surface, and we learned that consumer GPS devices supply satisfactory results. The next two chapters can be seen as an excursion into digital maps and navigable networks, which are needed to explore any area on the globe with a real object. This chapter will guide the reader to process OSM deliveries to significantly reduce map data. The next chapter will show how to process this data into a cartographic part, a network part, and an administrative part from one map source.

We will go through a complete cycle of processing digital maps with a simple map compiler in the `resources` folder: The steps are as follows:

1. *Collecting map details.* The first step to create a digital map is the collection of *geometry* data in the field, as demonstrated with the

79

recorded traces in the small application `MovingObjects` in the preceding chapter. The second major building block is the *administration*, which describes the political structure of a country and can hardly be determined in the field.

2. *Creating maps.* After collecting individual traces, field workers can make use of *graphical front ends* for analysis and modification. The traces are simplified and validated against other sources, like satellite pictures, elevation contours, and existing map data. The geometry is then verified and the points are drastically reduced and labeled with features and attributes describing the real world.

   Making digital maps is a never-ending process as the world changes constantly. Accordingly, each digital mapping organization has one main server to hold *the* map. Field workers "check out" the area of their responsibility, add new content, remove old content, or modify the data, before transferring data back into the *live map*.

3. *Delivering maps.* The main map server (or mainframe) runs around the clock to validate the incoming data and to enforce the (company's) *map specifications*. In given intervals, different parts of the map are extracted from the live map into predefined *map coverage* and *formats*. Afterward maps can be enriched with additional (third party) data.

4. *Processing maps.* Digital maps are gigantic chunks of geocoded data and the mapping company's live maps are designed to hold almost any data collected by field teams. It is important to understand that digital map deliveries *are not* customized for applications like navigation systems or online maps. It is necessary to *process* the map deliveries to fit various applications.

   This process from map deliveries to a *physical storage format* (PSF) is usually referred to as *map compilation*. Map compilers clip data, reorganize data for fast access with low memory consumption, and separate data for rendering, navigation, etc. Each map compiler is build to fit the requirements of the mapping application. While creating maps is a manual effort of many people, processing maps is automated.

5. *Using maps.* After the vendor of a navigation system, for example, has created a PSF for his system, the system can "read" the map on a physical storage for navigation. A navigation system is completely useless without a digital map and the system can only work with details (features, attributes) available on the map.

7.3   A Map Compiler for OSM Data

In what follows here, we approach the problem as field workers and map engineers. A basic structure of an OpenStreetMap compiler is provided in the resources folder `OSM.compiler`:

```
../resources/OSM.compiler
                +-- deliveries
                +-- products
                +-- tools
                    +-- josm
                    +-- Kosmos
                    +-- LinkCompiler
                    +-- osmosis
                    +-- OSMparser
```

You should move this folder to a large hard disk, where you can place your downloaded (input) and processed map data (output). The batch files were created for Windows systems and can be replaced with Unix shells. The main goal is to make each compiler step accessible via a command line. This way, the steps can be invoked from a higher-level script to wrap the entire production process in one place—later. The `tools` folder holds third-party tools; you should download the latest versions for best results. Of course, you should use the opportunity to get familiar with `openstreetmap.org` and the related webpages and tools.

As you will see, a map compiler has to leverage the usage of disk space and processing time. Therefore, the text is based on the production of the projects in the `resources/gps` folder, and each sample compilation is described with statistics for benchmarking. It is suggested that you follow each step with the concrete datasets and then repeat the step by processing data for your own area and project. By the end of Chapter 8, your compiler will be set up, and your products (physical storage formats (PSF)) should be (pre)defined. From there on, the entire process for all defined products can be fine-tuned and scheduled, logfiles can be parsed for fast failure, versioning can be added, etc.

Each section/processing step provides a brief overview of the mapping required, followed by concrete instructions on what is to be done and supporting statistics for design decisions to develop a strategy.

Warning! The map compilation of PSFs is a one-way process. The results can heavily deviate from the map source format and you should make sure *not* to upload them to the OSM server!

7.4   Collecting Geometry in the Field

The mapping scenario `MovingObjects` represents the modern approach to making maps. If you simply watch the traces spread over the screen (without the map image), you can watch the map drawing itself. The availability of GPS devices allows anyone to collect map data at low cost and with high accuracy. Every navigation system and many smartphones collect data all the time—and the manufactures know how to retrieve them. This is changing the mapping business dramatically as it doesn't require geo- and cartographic skills.

Commercial vendors of digital maps have large field teams all over the globe, driving (tens of) thousands of kilometers every month to collect every piece of information that might be of interest. Since every field trip consumes manpower, time, and money, the field cars are equipped with multiple sensors from differential GPS to video cameras. For best accuracy, the measurements of the GPS are compared to the traveled distance of the car and each change of heading is marked. Fieldwork has to consider the difference of the distance traveled between left and right tires, tire pressure, wheel slip, temperature, etc.

Due to different time zones and many other factors, it makes sense to always have *the* map available online for the field team. Map engineers can check-out (download) any part of the map, edit it, and finally check it back in (upload). Large teams have to be synchronized, concurrent editing has to be reintegrated, and precise timing schedules exist, when a map is locked and flushed to snapshots and extracted to different formats in consecutive steps.

The geometry represents the fundamental grid for any map and the majority of it needs to be collected only once. However, the collected traces can not be integrated directly into a map. Besides deviating slightly from the real road geometry, there are simply too many positions. The scenario RGB-BUELL includes about 50 traces with about 40,000 points. Each trace can be loaded and compared against the OpenStreetMap data with the Java OpenStreetMap Editor (JOSM).

The Java OpenStreetMap editor. In order to get your own graphical mapping front end, please visit

    josm.openstreetmap.de

and download the latest JOSM version[1] to the following directory:

    ../resources/OSM.compiler/tools/josm<version>

JOSM is a graphical front end to view digital maps in the OpenStreetMap (XML) format and GPX traces. Once you have downloaded JOSM and opened the editor, you can load the GPX files of the RGB-BUELL-NW scenario.[2]

---

[1]For this chapter version 3592 2010-10-05 was used.

[2]Right click on the GPX layer and convert to data layer to see a consecutive trace.

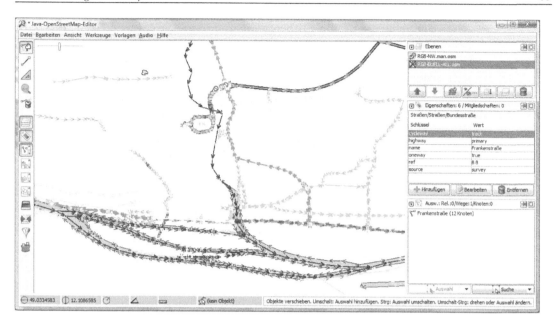

Figure 7.1. JOSM is the front end for manual OpenStreetMap data attribution and editing.

Next you (the fieldworker) can download the existing data of the area covered by the traces. Go to the File menu and click "download from OSM server..." and enter:

```
minlat = 49.00 minlon = 11.90
maxlat = 49.12 maxlon = 12.15
```

This bounding box will download about 7 MB and you should see a very colorful screen display after a few minutes (see Figure 7.2). What you see is the completed work of other field workers in this area. Possibly the size of the bounds needs to be iterated, since the (growing) amount of data varies for different regions. Before doing anything else, you should save the data as an .osm file like[3]

```
../OSM.compiler/deliveries/RGB-NW-JOSM.osm
../OSM.compiler/deliveries/<yourProject>-JOSM.osm
```

With JOSM you can look at the traces, zoom to each point and visually compare the traces against existing streets. Each node of a road indicates a *maneuver point* or a change of direction (*shape point*). The visual tool

---

[3]Add -JOSM to the file name as a reference for another file produced on page 89.

can make use of many other geospatial data sources to be compared with the traces: satellite pictures, blueprints of buildings, parks, etc.

This example indicates the restrictions of the graphical front end. For one thing, the map area is not very large compared to an entire country. Second, the word "graphical" implies human scanning and handcrafting by carefully validating every trace against information previously stored. This step relies completely on the human eye and can hardly ever be automated. Maps are always hand made—which makes them expensive. On the other hand, geospatial data is readily available, and there is no doubt that sooner or later the globe's surface will be captured with centimeter precision (in relevant places).

7.5   Map Attribution with JOSM

Generally speaking, the nodes (2D and 3D geopoints), links (edges, lines), and faces (polygons) represent the *only* fundamental primitives of a digital map and are sometimes referred to as level 0. For example, a digital map does not provide any curvature, and the map user has to apply spline algorithms, etc., for smooth map displays.

After matching the traces to satellite pictures and existing data, the next step is to manually add real-world significance—level 1. In JOSM you can zoom into the map you have downloaded, use the selection tool, and highlight any object with a mouse click. The properties pane will show you the "meaning" of the object.

Click on a road link to see features like

```
highway: primary
   name: Frankenstrasse
 oneway: yes
    ref: B 8
```

To study the main OSM features go to

```
wiki.openstreetmap.org/index.php/Map_Features
```

As the name indicates, openstreetmap is focused on streets to provide maps that can be used to navigate vehicles. Naturally, street maps are the most sought after, but they are pretty useless for boats and airplanes. On the other hand, navigational devices have found their way from cars into peoples' pockets and it is a natural extension to add ways for walking, biking, etc. It is only a matter of time until the road grid is captured for all industrial countries, where cars are an integral part of every day life.

When the method of delivery was a CD, individual maps were produced for single countries, and the driver had to change the CD when traversing a country border. Today, most navigation systems are equipped with continental maps and one can navigate from any house in Portugal to any house in Moscow, for example. This requires a map to hold *a navigable road network* connecting to every single address on the map (or continent).

Digital street maps have more to offer than geometry and administrative coding—they include traffic rules.[4] Observing traffic is a main reason for gathering map data in the field. Field workers have to be capable of recognizing complicated or dangerous driving situations and add important guidance assistance to the map data.

After the geometry (level 0) is in place, the field team researches for relevant information to be added to the map. Each map element is digitized with meaningful attributes (level 1). The sources for road names and house numbers range from paper maps to phone calls with local authorities. Usually field offices are located in their region of responsibility to ensure as much local knowledge as possible. The graphical front ends are used to control the accuracy and position of each node, to move them to the centerline of a road, and so on.

The heart of a digital map is the *map specification*. The specs describe how to code every single piece of map data. Global specifications enable customers to process each map regardless of coverage and cultural differences (i.e., driving side). Map data can be processed in one tool chain to produce general functionality to specific guidance in complex situations (level 2).

A strict specification is the prerequisite for quality assurance and batch automation to enforce the map's consistency worldwide. While the OSM features and attributes are kept rather general, due to its open source nature, commercial map specifications can easily cover more than a thousand pages, and a single link can have more than a hundred attributes attached to it.

Typically street maps offer *content* with

- geometry: links, nodes, elevation, connectivity;

- administrative hierarchies: county, state, city, postal areas, zones;

- cartography: railways, waterways, lakes, woodland;

- POIs: well-known places (instead of formal addresses);

- validated navigable networks for different classes of roads;

---

[4]In the context of the ROAF vision *rules* are guidelines and can be violated, *laws* can not.

with *attributes* to describe

- roads: name, surface, width, lanes, accessibility, house numbering;

- restrictions: gates, turns, times, vehicles, speed;

- time dependencies: reversible one-ways, seasonal closure of roads;

- signs: text, exit and route numbers;

- multiple languages and voice phonemes (in various pronunciations)

and continually growing *coverage* (and information density) of America, Australia, Europe, South Africa, Middle East, and other parts of the world.

7.6 Map Formats

The map specifications describe the content of a digital map. They *do not* describe the proprietary structure of the company's map server, nor its database. The internal database is optimized for concurrent map production, field worker's (audio) notes, and validation and must be flexible for modifications. Externally map suppliers should comply with international standards to deliver navigable databases that can be used with third-party tools. Different standards exist for different kinds of applications for desktop, server-based, or GIS navigation and routing. The formats differ with different content, coding, and merging abilities depending on the customer's business. After different portions of the map have been quality ensured, they can be copied from the live database and extracted to widely available data formats processed by system manufacturers.

One of the most prevalent formats in the industry is GDF 3.0,[5] GDF incorporates a wide variety of features and is an international (slowly fading out) standard for delivering navigable databases. Just like in OSM files, GDF begins with a list of *all* nodes of the map (level 0). The information to link the data is spread across the rest of the file using IDs as pointers. This structure is extremely inefficient for direct access by an application. Every system uses different (map) features and every vendor uses a proprietary data schema filtered and optimized for fast access (routing). The process of converting map deliveries to a physical storage format of the target system is called *map compilation*.

You will learn how to build your own *map compiler* to create dedicated maps from `openstreetmap` for your own application (ROApp). The process will give you a good idea of how time- and resource-consuming the process is. Professional map compilers use the largest storage and fastest machines

---

[5]GDF = Geographic Data File, based on CEN (European) Standard.

available. Although today's personal computers are adequate to handle large map files, you will find it easy to keep your CPU busy. OSM extracts the map into various files in given intervals (release schedule); they can be downloaded at

    downloads.cloudmade.com

The goal in setting up your map compiler is to achieve an automated compilation from the delivery of well-known map formats to the target, the physical storage format and the final product for a map application. First, you should define the coverage of your product; then you need to develop a strategy for the map compilation.

For example, if you want to create a product for the major roads of Europe, which is represented by much less than one percent of the map data, it is suggested to load the full continent and then strip it down in the compilation process. This approach involves large amounts of map data, disk space, and processing time.

Still, it is much easier to reduce a given map than merging smaller maps. In order to merge with neighboring countries, the map compiler has to be able to detect redundant data at the country borders and get rid of it. Also, the country files have to have identical ID spaces at the borders, since a geometric analysis is always time consuming and not necessarily deterministic.

So, the first decision is the starting point for the map image in the RGB-BUELL-NW scenario. For a large number of similar scenarios in Europe, it would make sense to prepare the map data for Europe and then clip each product from there. In the context of this book, the compilations should be executable on a regular PC in reasonable times.

Table 7.1 provides a first estimate of disk space and download times. Since the download is already part of the production process, five and a half hours is too long to get started. Knowing that the European product was about 750 MB in August 2008 makes it clear that the strategy might need to be changed over time. The zipped Germany file consists of almost one quarter (23%) of Europe, which is possibly not obvious when

Coverage	Zipped file size[a]	Download time[b]	File name
North America	4,110 MB	4 h 30 min	north.america.osm.bz2
Europe	4,967 MB	5 h 30 min	europe.osm.bz2
Germany	1,151 MB	1 h 15 min	germany.osm.bz2

[a] in November 2010
[b] assumes average download rate of 250 kB per second

Table 7.1. File sizes and download times for various OSM deliveries.

considering the geographical size. Since Germany has a large car industry and a dense population, it is one of the biggest countries with respect to street kilometers. Therefore, it serves as a useful intermediate benchmark for country and continental file sizes.

7.7   Processing Maps with Osmosis

After downloading a continental and/or country openstreetmap, you should visit

`wiki.openstreetmap.org/wiki/Osmosis`

and download the `OSMosis` tool to a `../tools/osmosis-<version>` folder.[6]

Osmosis is a command-line Java application to process OSM data in multiple ways by applying the SAX parser on OSM XML files. Since this book only covered the DOM parser earlier, you might want to have a look at the Java sources to get familiar with SAX functionality.

(Un)Zipping Maps

With the command line (without linebreak!), the file is unpacked:

```
D:\<full path>\tools\osmosis-0.37\bin\osmosis.bat
          --rx germany.osm.bz2 --wx germany.osm
```

Although osmosis is capable of processing compressed files, other tools operate better on uncompressed data and tools like JOSM can only load (limited) uncompressed data. After processing a digital map, it should probably be recompressed for shipping with the following command line options:

```
--rx germany.osm --wx germany.rezip.osm.bz2
```

So it's a good idea to get some more precise numbers for the timing for zipping and unzipping (see Table 7.2).

Note that the built-in bzip of osmosis does not produce optimal unpacking or packing results, and the duration is not linear with the file sizes. In the long run, it pays off to configure an exclusive packer for the platform's native implementation. This is vital for commercial maps, where the Europe map exceeds 100 GB and it can easily take two days only to get the dataset ready for processing. For the time being, osmosis will be used as a multipurpose tool.

Another important aspect of commercial maps is the checksum of a delivery, which is important for legal claims, when tracing map errors back

---

[6]For this chapter version `osmosis-0.37` was used.

Coverage	Zipped	Unzipped	Ratio	Unzip	Rezip	Ratio
Germany	1,207 MB	14,875 MB	8%	2 h 30 min	5 h 09 min	200%
RGB-NW	480 KB	5,950 MB	8%	5 min	10 min	200%

Table 7.2. Uncompressing and recompressing OSM deliveries. (Hardware: dual core CPU 1.67 GHz, 4 GB RAM, Win32 OS.)

to the source. When using different tools (or even one tool) unzipping and rezipping a file most likely does not produce the identical byte size.

The benchmarks for the Germany map indicate that it takes almost four hours to download the map and unpack it. For the single RGB-NW project this is a lot of pre-processing. In a larger context, it might be worthwhile to download and uncompress an entire continent for a whole day in order to clip a few hundred bounding boxes all over Europe for one application in different locations.

Clipping Maps

Compressing is not really a map-related process and does not require osmosis. Vice versa, osmosis can execute map-related functions on packed OSM data. Earlier JOSM was used to download and save a bounding box directly with the front-end tool. With osmosis, another approach can be chosen to get the same map area. With the command line options below, the RGB-NW area can be clipped directly from the packed Germany map.

```
--rx germany.osm.bz2
--bb left="11.95" right="12.10" top="49.10" bottom="49.02"
--wx RGB-NW.osm.bz2
```

By unpacking it (see Table 7.2), the file should nearly match the file saved directly from JOSM on page 83.

With the command line options, the RGB-NW area can be clipped directly from the unpacked Germany map.

```
--rx germany.osm
--bb left="11.95" right="12.10" top="49.10" bottom="49.02"
--wx RGB-NW.osm
```

So there are different ways to get to the RGB-NW.osm file:

```
germany.osm.bz2 - 2h 30' - germany.osm
       |                         |
     2h 05'                     20'
       |                         |
  RGB-NW.osm.bz2 - 5 sec  -  RGB-NW.osm
```

For this one project, the choice of a path doesn't make a big difference. For many cities in Germany, it does make a difference, since the processing of different products can be parallelized and clipping from unpacked files is obviously faster.

In the end, all the processing time and disk space should be logged with every map compilation and for every change of the setup; even for every option applied to osmosis. A map compiler does not scale linearly with changes and can always be fine-tuned. Also, the storage management and moving of files over the network takes time. In the case of the OSM compiler, it would be helpful to abstract the deliveries folder from the hardware (symbolic links). This way the compiler can work with one structure, regardless of the underlying hard disks.

Extracting Map Layers

When downloading via JOSM, or clipping with osmosis, the result set includes *all* map data available in the given bounding box. A map compiler processes this given map data to dedicated maps for a concrete application. Therefore, it is vital to separate the data for different purposes. For example a navigation system usually requires one map for rendering and another for navigation. Osmosis is capable of extracting different *layers* of a map. These layers are basically defined by the attribution of the map primitives. With the command-line options below, osmosis extracts the primary highways from the input map to a separate file:

```
--rx RGB-NW.osm
--wkv keyValueList="highway.primary"
--un
--wx RGB-NW_hw.1.osm
```

Table 7.3 lists the extraction times and resulting file sizes, when splitting the RGB-NW.osm file into different layers. The whole process only takes 30 seconds. Note that the sum of the layer maps is about two and one-half times as big as the source file. Consequently, this compilation step requires about four times the size of the source to avoid hardware failures.

In case of RGB-NW each road layer can be visually compared to the GPX traces[7] with JOSM and adopted accordingly to match the Moving Objects with the map. For the simple project RGB-NW, every layer was compared to the traces and manually cleaned and then merged back together to one OSM file with JOSM.

---

[7] I.e., convert RGB-BUELL-NW-88-080523.gpx into a layer for a better visualization.

Map feature	Extraction (ms)	File name	File size
places	2496	RGB-NW_place.osm	4,363,790
highway.trunk	2303	RGB-NW_hw.tr.osm	917,613
highway.primary	2334	RGB-NW_hw.1.osm	976,845
highway.secondary	2289	RGB-NW_hw.2.osm	930,736
highway.tertiary	2529	RGB-NW_hw.3.osm	1,099,032
highway.unclassified	2369	RGB-NW_hw.un.osm	1,070,351
highway.road	2655	RGB-NW_hw.rd.osm	958,407
highway.track	2696	RGB-NW_hw.tk.osm	1,669,279
waterway.river	2565	RGB-NW_ww.rv.osm	957,725
railway.rail	2430	RGB-NW_rw.rl.osm	933,845
natural.wood	2991	RGB-NW_na.wd.osm	970,275
natural.water	2412	RGB-NW_na.wt.osm	943,960
	30069		15,791,858
		RGB-NW.osm	6,092,583

Table 7.3. Extracting map layers from RGB-NW.

7.8 Parsing OSM Files with the `OSMparser`

JOSM and osmosis are part of the `openstreetmap` map-production process. The tools can be used to check-out map data, modify it, and check it back in to the map server. Due to the large number of users and contributors, these standard tools should be used when possible. On the other hand, XML is also a well-defined standard and OSM files have a simple structure:

> `wiki.openstreetmap.org/wiki/Elements`
>
> Our maps are made up of only a few simple elements, namely nodes, ways, and relations. Each element may have an arbitrary number of properties (a.k.a. Tags) which are Key-Value pairs (e.g., highway=primary).
>
> A tag is not an element but a property attached to a node, way or relation.

A look at the file `RGB-NW.osm` (OSM version 0.5) with a text editor reveals a lot of map-production-related information:

```
1: <node id="232389894" uid="109925" changeset="2150422"
        user="hajopei" timestamp="2009-08-15T08:35:28Z"
        version="2" lat='49.0455424' lon='12.13113'>
2:    <tag k='created_by' v='JOSM' />
3: </node>
```

Most of this information is irrelevant for map productions, and the node element could be reduced to one line:

```
<node id='232389894' lat='49.0455424' lon='12.13113'/>
```

This data reduction from 197 to 59 bytes[8] (30%) seems promising enough to use it for data delivery to save a lot of disk space and processing time for subsequent steps of the map compiler.

Program Design

Generally speaking, all processing steps of a map compiler are command lines for the operating system. Therefore, many OS tools can be adapted for different processes. A typical tool is the stream editor to perform text manipulation on an input stream and write the output stream to a file. While the stream editor is a rather generic tool, it can be helpful for a map-compilation team to have a command line tool more sensitive to OSM files.

OSMparser.java

> For a deeper understanding of the parser for OSM files the reader should open the file OSMparser.java and set the command line arguments in the IDE. In this way, the parsing can be repeated multiple times and stop points can be set in debug mode to analyze the temporary data structure. For example, in Eclipse you would open
> set the Run Configurations > Arguments > Program arguments: to enter the following example (in one line without breaks)
> -rx "D:<yourpath>\products\GER\osm\gui\RGB-NW.osm"
> -wx "D:<yourpath>\products\GER\osm\gui\RGB-NW.xWsHW.osm"
> -xways "<tag k=.highway.*/>"

The OSMparser uses the idea of a streaming editor to read an OSM file stream, parse, and modify elements, and write a modified OSM stream back to the file system. The main processing steps are to

1. read every line of the OSM UTF8 input file;

2. identify one of the three OSM element types: <node|way|relation>

3. create a List<String> element for manipulation;

4. delete lines from each element matching a regular expression;

5. delete or filter elements identified by a regular expression;

6. write the modified elements to a specified output file.

---

[8]Note that one integer and two double values require about 20 bytes in a database.

The listing implies two additional conditions for the OSM delivery. For one thing, it is important to use UTF-8 capable tools to write OSM files in order not to break the unicodes and corrupt the file. Also, the OSM delivery should be consistent with a well-formed XML file *and* every line needs opening and closing tags.

The main method of the parser is `parseOSM()` . First two piped UTF-8 streams are created:

```
BufferedReader reader =
    new BufferedReader(
        new InputStreamReader(
            new FileInputStream (  inFile ), "UTF8"));

BufferedWriter writer = ...
```

Then each line is read until the end of file is reached:

```
while ((line = reader.readLine()) != null)
{
// identify as element
    element = createElement( elementType, line );
    element = processElement( element );
// write element to output file
}
```

The method `processElement( element )` returns a (manipulated) element or `null`, if the element should be removed. Different methods to manipulate the elements are used according to the command-line switches.

These methods may serve as templates and look like:

```
List<String> methodName( List<String> element )
```

The first method `cleanAttributes( element )` does exactly the data reduction described with the `<node>` element. The redundant attributes are hard coded:

```
String[] attributes = {"timestamp", "visible", "user",
            "action", "version", "uid", "changeset" };
```

The second method `removeLine( element, regEx )` introduces a second parameter: a regular expression. A complete line is removed from the element if it matches a regular expression. Additionally, the method optimizes the element, while conserving XML validity.

The third method `boolean identifyElement(element, regEx )` can be used to check the entire element for a regular expression. The Boolean result can be used to keep or dispose of the element according to the command-line switch.

Reading regular expressions from the command line makes the tool very flexible for all kinds of situations that a map engineer might encounter.

THE JAVA TUTORIAL > ESSENTIAL CLASSES > REGULAR EXPRESSIONS

> This lesson explains how to use the `java.util.regex` API for pattern matching with regular expressions. . . . Regular expressions are a way to describe a set of strings based on common characteristics shared by each string in the set. They can be used to search, edit, or manipulate text and data.

Program Usage

The `OSMparser.java` file can be found in the default package of the `book/src` folder, and the compiled `class` file can be copied to any folder and invoked as follows:

```
java OSMparser -rx path\in.osm -wx path\out.osm ...
  1: ... -clean
     clean file from attributes and lines not of interest.
  2: ... -xlines     regEx
     delete all    lines matching the regular expression.
  3: ... -xelements regEx      (complement to -felements)
     delete all elements matching the regular expression.
  4: ... -felements regEx      (complement to -xelements)
     filter all elements matching the regular expression.
  5: ... -xways      regEx
     delete all ways matching the regular expression.
```

With the command line below, the delivery unzipped in two and one-half hours (see 88) can actually be reduced by 49%—to 137 minutes. Now you can see why batch automation is vital. The process of downloading, unzipping, and cleaning for a country like Germany already takes about six hours (see Table 7.4).

```
java OSMparser -rx germany.osm -wx germany.clean.osm -clean
```

The program collects some statistics on the fly and provides a short overview:

```
    read: germany.osm
   write: germany.clean.osm
starting '-clean' at Sat Nov 13 14:20:00 CET 2010
start parsing germany.osm
statistics  germany.osm <> germany.clean.osm
    lines: 209718750 <> 198429239
    nodes: 64721366 <> 64721366
     ways: 8929165 <> 8929165
relations: 134531 <> 134531
  ending '-clean' at Sat Nov 13 16:36:26 CET 2010
```

With the command line (without linebreaks)

```
java OSMparser -rx germany.clean.osm -wx germany.norel.osm
          -xelements "<relation id=[\"']"
```

all relations can be removed from the file, since they will not be used with this map compiler.

> The batch file `../deliveries/processDelivery.bat` lists different processing options of the parser. The batch should be adopted to the needs of your own compiler.

A nice setup to test regular expressions is an open command line box to repeatedly call the parser with different expressions. A smaller file (`RGB-NW.clean.osm`) is created in a few seconds and can be opened with a (UTF-8) text editor (or diff tool) after the first run. Consecutive calls of the command line will write over an existing file and the output can be updated in the editor window. Note that the overwriting of preexisting files could destroy an earlier parsing. Nevertheless, the tool should not (have to) wait for a keyboard input, since this could interrupt a longer compilation.

Note that

- the input file has to have the extension `.osm`;

- Java splits the command line arguments by the space. Thus any string (and regex) with a space should be masked with `""`;

- the regular expression provided via the command line is compiled and immediately returns a hint to the command line, if it is invalid.

  ```
  Please review regular expression:
  Unclosed character class near index 15
  <relation id=["'
                 ^
  ```

7.9 Toolchain Configuration

The `OSMparser` introduces regular expressions as a powerful tool to manipulate OSM data. However, it is kind of numb to map semantics and can easily make the output file unusable. For example, the `-felements` switch can be used to fetch distinct elements and analyze them in an editor. This can be helpful for concrete problems, but the output file is useless for further processing.

When setting up a map compiler, the engineer has to be aware of the side effects of every tool used. To make sure that the `*.clean.osm` files are still valid, they can be repacked after cleaning to include unicode, XML, and, possibly, OSM validation. Although this process takes some time, it secures the production and, in the long run, it's also a good idea to archive the map sources. This *toolchain harmonization* can implicitly detect unknown problems and they can be fixed as they occur. Each tool and each map file creates a new version and every change can cause surprises.

When trying to zip the cleaned file with osmosis

```
D:\<full path>\tools\osmosis-0.37\bin\osmosis.bat
     --rx germany.clean.osm --wx germany.osm.bz2
```

one of these surprises pops up:

```
org.openstreetmap.osmosis.core.OsmosisRuntimeException:
Node 104936 does not have a version attribute as
OSM 0.6 are required to have. Is this a 0.5 file?
```

The exception describes the problem: the parser was initially developed with OSM 0.5 files and since the requirements have changed, the tools need to be adapted. After this exception from osmosis, the OSMparser has to be modified and the delivery has to be processed again.

This demonstrates the necessity to test a compilation with a smaller file through the full chain *before* using it on really big files. After sufficient understanding and testing (development), you can run a number of parsing steps on a full country file (testing) and then integrate the steps in your map compiler (production).

Note that a new switch was added to the parser: with -clean, the file is cleaned from unnecessary OSM attributes, while -cleanAll would remove all attributes. The implementor needs to be aware that the -clean switch is already part of the map compiler and that a new behavior could affect all products. The OSMparser is *not* a tool for map *production* like osmosis. So the latter option (-cleanAll) should only be used after the file has passed all OSM tools. With every change in the toolchain, the map engineer has to check backward compatibility issues. Take some time to load a RGB-NW.osm file into JOSM, modify the map, and observe the changes in a text editor or with a diff tool.

Batch Processing

The simple processing chain takes eight to ten hours to get the Germany file in place, and the consumed disk space is a factor 23 larger than the initial delivery. The CPU load is high during the process, and it might be a good idea to include cooling pauses, depending on the hardware used (see Table 7.4).

The parser is just a small, yet effective, tool that a map engineer can use in a map compilation and easily add more functionality, for example, a -stats switch (see page 107) to collect statistics. Or it can serve as a

File name	Process step	Ratio	Time	File size
germany.osm.bz2	download		1h 15 min	1,207,301,173
germany.osm	uncompress	8%	2h 30 min	14,875,159,473
germany.clean.osm	clean	73%	1h 52 min	10,910,796,283
germany.clean.osm.bz2	recompress	89%	3h 54 min	1,081,087,220
			9h 31 min	28,074,344,149

Table 7.4. Preprocessing Germany for a map compilation.

template for a completely different task like the `LinkCompiler` introduced in Section 8.3.2 .

The batch file `OSM.compiler/deliveries/processDelivery.bat` wraps up the preprocessing stage. You should interactively test different regular expressions on a smaller file and remove information not needed for the final product to reduce the size.[9] The implicitly-called batch file `OSM.compiler/tools/tools.bat` is used to configure all tools for a compilation. The next development cycle to improve the compilation for early error detection adds log files for every command line and another process would search through these files to look for error messages.

Another thing is obvious: there are now two versions of the `OSMparser` for different OSM file types. Hence, the map compiler needs a versioning system. In practice, the `OSMparser.java` file can be placed in its tools folder, and the batch files should compile the file on the fly (with the `javac` command line). The compiler should be labeled for every production run, since products usually have to be supported for at least one decade. (The challenges of forward and backward compatibility are great.) The OSM files should *not* be checked into the versioning system. Instead, they should be renamed with the time of delivery (or download) and archived for a potential re-production.

7.10   Rendering OpenStreetMaps with Kosmos

After processing the file Germany, clipping `RGB-NW`, and splitting it into layers, the result is similar to the data downloaded with JOSM. The sample compiler was created to give an idea of map processing and to show ways to process much bigger bounding boxes than JOSM can download.

Before extending `RGB-NW` to `RGB` have a look at the result file that was manually split, cleaned, and merged back to one file of about 500 KB:

    ../products/RGB-NW/osm/RGB-NW2010.osm

---

[9]In the next chapter, the Germany file will be reduced significantly.

When opening the file with JOSM, the map image from `MovingObjects` still can't be seen. Of course JOSM shows a map, but not in a format suited for consumers. JOSM is a support tool for map engineers and doesn't waste time to calculate projections. Most consumer applications, like navigation systems or maps on the web, have a front end which displays the map. Although the map geometry might come from one source, the appearances can differ completely.

Rendering Maps

In order to render `RGB-NW2010.osm` another OSM tool will be introduced.

> Go to `wiki.openstreetmap.org/wiki/Kosmos` and download the Kosmos tool to a `../tools/Kosmos-<version>` folder. (For this chapter, version Kosmos-2.5.405.6 was used.) Don't forget to adapt the `tools.bat` accordingly!

`wiki.openstreetmap.org/wiki/Kosmos`

Note that Kosmos is no longer actively maintained and you should consider using the successor Maperitive. Kosmos was primarily designed to be used by OSM users on their own computers to

- render OSM maps interactively,
- print OSM maps,
- set up a local tile map server,
- use their own Map rendering rules or share rules stored in OSM Wiki pages,
- perform certain tasks from the command line.

Kosmos was designed to be as simple to set up as possible, without losing much of rendering capabilities.

Kosmos was chosen here as a lightweight OpenStreetMap rendering platform. After launching the Kosmos GUI, you can create a new project and load the file `RGB-NW2010.osm` to finally see the familiar image (see Figure 7.2). The image file can be created with the menu: File - Export to Bitmap. Before going to the next step of automation, you should get familiar with the tool and modify the map appearance.

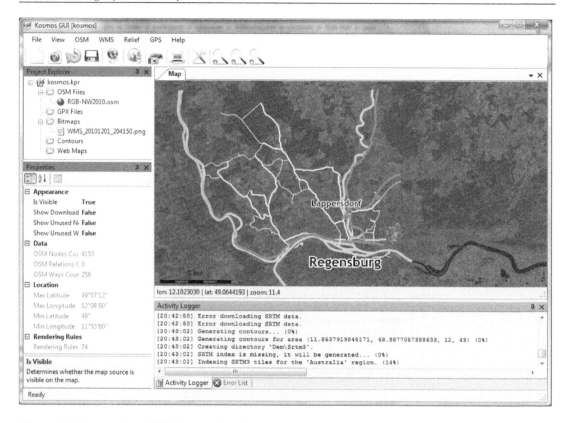

Figure 7.2. Kosmos is a GUI front-end and command-line tool to render Open-StreetMap data.

Creating Map Tiles

As stated above, Kosmos is a command-line tool. Before automating rendering, a graphic designer has to define the rules, and this can only be done with a GUI front end. The following steps have to be executed to prepare the rendering for the next RGB project:

1. Open a command line box, go to `../products/RGB/` and enter `createRGB`

   (a) the `germany.clean.osm` file has to reside in the `deliveries` folder;

   (b) the subfolder `osm` has to exist and existing files will be over written;

   (c) make sure the last line (`%kosmos% ...`) in `createRGB.bat` is commented out;

The process takes about 20 minutes (on the known hardware reference) and the subfolder `osm` should be filled with OSM files similar to Table 7.3.

2. Open the Kosmos GUI: `../tools/Kosmos-2.5.405.6/Gui/Kosmos.Gui.exe`

   (a) open the Kosmos project file `../products/RGB/kosm/RGB.kpr`

The Kosmos GUI lists all expected project files in the "Project Explorer." The file names are grayed and the "Error List" shows problems encountered when loading the files. Right click on the folder "OSM Files" and open all the layers compiled with `createRGB.bat`.[10] Or open the files one-by-one to watch the drawing of each layer.

Now click on `RGB.kpr` in the "Project Explorer." The "Properties" window shows a single line "Rendering Rules." Enter the full absolute path to the file `../kosm/RGBrules.txt`. Save and refresh the project.

The file `RGBrules.txt` defines the rendering rules. Open the file in a text editor and look for the (single) line:

```
| River || {{IconWay}} || {{tag|waterway|river}} |
| Polyline (MinZoom=5, Color=lightblue,
  Width=5:1;10:2;17:14, Curved=true) || ||
```

Change the color `lightblue` to `black`, refresh the project, and look for black rivers.

Take some time to go through the "WMS" and "Relief" menus. You will easily find out how to add elevation contours, slope shading, a shaded relief, or a satellite image. These features provide instant feedback about the precision of the map. The file `../RGB/kosm/RGB.png` provides a possible appearance. Now, save the project.

The project XML file `RGB.kpr` consists of the above steps. This process could be automated by putting a `template.kpr` in the map compiler and filling in paths and filenames via a stream editor. The problem with that is that the paths have to be absolute in the project file and can not simply be moved to another directory with different paths.

After this manual design process, the project can be used for the automated production of map tiles with a (single) command line like[11]

```
D:\tools\Kosmos-2.5.405.6\Console\Kosmos.Console.exe
   tilegen D:\products\RGB\kosm\RGB.kpr
   48.89 11.58 49.4 12.57 8 12 -ts Tiles
```

Note that the batch file is preset to render the area clipped for RGB with the zoom levels 8 through 12. The duration of the tile rendering depends on

---

[10]Do not add the source file `RGB.osm`!

[11]See last line of `createRGB.bat`.

the details of the map, and it is a good idea to test with a limited number of zoom levels.

A small area like RGB can produce a lot of data:

```
'Tiles' folder:
            1,069 Folders
          235,923 Files
    2,700,925,950 Bytes    ~ 2.5 gig
```

The time consumption for eighteen layers can be significant:

```
level 0-15: 0:56  - 1:19  =    23 minutes !
level   16: 1:19  - 2:30  = 1h 11
level   17: 2:30  - 7:30  = 5h 00
----------------------------------
   total:   0h56min - 7h30min = 6 and one half hours !!
```

After some experimenting, all osm files and the Tiles folder can be removed again and now createRGB.bat can run all the way. In the context of a map compiler, the creation of products like createRGB.bat and createRGB-NW.bat can be run in parallel, if the hardware allows this.

TileMapViewer. We now go back to the *two* applications MapViewer and MovingObjects described on page 76, and enable the viewer to display the tiles[12] created above.

The packaging of roaf.book.ro.MovingObjects.java, roafx.swingmap .MapViewer.java, and roaf.book.map.osm.TileMapViewer.java is important in the larger project context:

- MovingObjects is real-object specific;

- the MapViewer is an external observer to real objects;

- the TileMapViewer is an OSM-specific add-on.

The implementation of the TileMapServer is straightforward:

1. The menuTileManger() instantiates a new TileMapServer(tilesDir) to read the tile directory.

2. The TileMapServer goes through the zoomlevels 0 to 17 to createTiled Maps for each existing directory.

3. The constructor of each TiledMap iterates through all subdirectories to allocate the images according to the vertical and horizontal matrix.

4. Since all tiles have the same pixel size only the very first image is loaded to get the horizontal and vertical pixel lengths.

---

[12]Find the tiles in ../resources/gps/RGB-BUELL/Tiles.

5. The method `loadMapArea` loads a bounding box for every `TiledMap`.

6. After all maps and tiles are allocated the server is ready to serve any of the maps with their `Tiles`.

7. A `Tile` is a recordset of an image (location) and its map area.

Section 6.5 covered steps *1*, *2*, and *3* of the quick start instructions `../deployed/MovingObjects.txt`. Please go through steps *1*, *2*, and *4* now.

Chapter 8

Making Maps Navigable

8.1 Introduction

After pre-processing digital map deliveries, this chapter will show how to divide the map into a cartographic, a network, and an administrative part.

The network represents the heart of a navigation system and it has to be converted into a format suited for effective routing. A `NavigableMap` is created to route from any point in the network to any other point chosen from a destination list.

In the next chapter a real object will be equipped with this navigable map to be able to sense its environment and to provide a means to *make a choice* where to go.

8.2 Map Compiler Branches

Processing a map delivery is the first step of a map compilation. Unnecessary information can be removed, while product-specific attributes remain in the optimized map file(s). The compiler generalizes the product portfolio on the front end and refines each product toward the back end. It has to be flexible and competitive and maintained and performance optimized, while ensuring backward compatibility using regression testing.

It would require an unreasonable amount of resources to build a compiler to exclusively process a Europe map, to extract Germany, and then to extract a Regensburg image for the `RGB` projects. After designing the processing chain for `RGB`, the compiler should be applicable to *all* cities and all products should work with the `MovingObjects` application.

A map is compiled according to the product coverage and feature set. For a navigation system, a map image is provided for the users' orientation and the moving vehicle is drawn on top of this map. The system does *not* depend on the projection of a map display. It works with the nodes of a map and is basically build to navigate along a given road network to a destination provided by the user.

Therefore, map compilers for navigation systems can be divided into three independent processes after cleaning and clipping a source map:

1. The *cartographic* branch creates a purely visual map—a number of image tiles; or tile information for on-board rendering. The visual map uses the full geometry with all available *shape points*.

2. The second branch creates a *navigable network*. The map data is reduced to *maneuver points* to build a *graph*. The network is validated to be *closed* to ensure that any node (origin) can be routed to any destination node.

3. The third branch compiles the *administrative* part of the map. Administration has to be organized in hierarchical structures for the user to search his destination with various approaches. Redundant character strings are removed to save space and addresses are connected to geographic positions.

It is vital that all three branches are compiled from one map source to avoid the end-user loosing his confidence in the system, when the moving vehicle is depicted driving across a field—or lake.

8.2.1 Slicing Germany

We can now reuse the data from the OSM compiler. The next project GER should minimize the entire map of Germany into small pieces for effective processing on personal computers or smartphones.

The project will be unfolded in

```
../OSM.compiler/products/GER
                          +-- admin
                          +-- carto
                          +-- networks
```

The cleaned osm file of Germany produced on page 96 represents the starting point of the project-specific compilation. The file should reside in the

```
../OSM.compiler/deliveries/germany.clean.osm
```

Processing a 10 GB file with osmosis and OSMparser takes hours and would take days to pass different branches to a final PSF. Today, many personal navigation devices (PND) can be updated via the internet and the end user probably doesn't want to download 10 GB. Some Europe PSFs are about 100 MB and can be loaded in one or two hours.

The next step is a product-specific data analysis. The batch file country.bat[1] was created to split the entire Germany map into different

---

[1]The name indicates that it can be copied, modified, and applied to any country.

files without loss. This file can be used to fine-tune the compiler tools to the underlying hardware.

The first step, to remove all relations in the file, can be used as a benchmark for the compiler and hardware. Osmosis requires more RAM than the OSMparser, and a choice should be made according to the platform used to get an idea of the maximum duration of a single conversion step. Note that the file extensions are hard-coded in the scripts and should follow a general naming schema. (This way the file names listed in the log files speak for themselves.) In case a project does require relations, this long process with little data reduction (1 - 2%) can be skipped. Then, the hard-coded file names in the batch `*.norel.osm` should also be replaced with `*.clean.osm`. Subfolders are hard-coded and their creation (`mkdir`) can be added.

Network layers. The three layers

```
highway.motorway   waterway.river   railway.rail
```

represent three different networks. The extraction of each network takes about 25 minutes[2] with the file sizes of 32 MB for highways, 67 MB for waterways, and 92 MB for railways. These networks were chosen since they are completely independent of each other and don't overlap. Cars, ships, and trains don't interfere with each other, while commercial vehicles, passenger cars, and pedestrians do share parts of one network.

Cartographic layers. The `GER` project aims to produce a map of Germany with a pixel size close to a typical computer screen. In order to get some green and blue color on the map (background), some `landuse` and `natural` features are extracted:

```
landuse: forest farm residential industrial
natural: wood water
```

Looking at the image of the Germany map for a computer screen, it is obvious that the forest polygons smaller than a few pixels could be removed. Nevertheless, this process is not straightforward with the tools we have introduced. The process would require a tool with *spatial* functionality to calculate the actual size of a polygon and remove it below a certain threshold. Also, the number of nodes could be significantly reduced without changing the low-resolution shapes.

The `OSMparser` could be improved to detect polygons, using methods to match first and last node. Then, the parser could dispose of all polygons that have a small number of nodes and every second node of the large polygons, etc., rerun as needed.

---

[2]Find the exact figures in `germany2010.log`.

Administrative information. The third major compilation step is the extraction of administrative information. Basically, most countries can be subdivided into an administrative hierarchy of city, county, and state, which is represented in the destination selection of any navigation system. In many commercial maps, every single link has a reference to the administrative section to enable fast detection.

The country script extracts a %1.cities.osm file with all *nodes* attributed with place.city. In OSM these nodes have the admin hierarchy codes as follows:

```
<node id="240120926" lat="49.0159295" lon="12.0956268">
  <tag k="place" v="city"/>
  <tag k="name"  v="Regensburg"/>
  <tag k="is_in" v="Regensburg,Oberpfalz,Bayern, ...
                ... Bundesrepublik Deutschland,Europe"/>
  <tag k="population" v="129323"/>
```

Then, the political borders are extracted to %1.borders.osm with all ways for drawing a political map together with the natural areas attributed with boundary.administrative.

Map algebra Especially when setting up a dedicated compiler for the first time, the engineer can only guess which features to pick for the final map. Therefore, the country script *can* be used to extract the entire landuse and natural layers for comparison. Note that this operation might require additional RAM for osmosis as applied in the script.

With simple algebra, estimations can be made:

```
landuse - ( forest + farm + residential + industrial ) = ...
natural - (   wood + water ) = ...
```

For exact figures, two additional command lines are added to the script to actually apply this algebra on the files and produce the following files:

```
germany.landuse.rest.osm  germany.natural.rest.osm
```

And, finally, the script can be used to remove extracted elements from the map:

```
Germany - landuse - natural - networks - cities
                      = germany.rest.osm
```

Note that this operation is divided into subsequent steps and osmosis needs a lot of RAM. Generally, all command lines pass through the complete input file to improve readability, while osmosis does offer pipelining. The reduction steps do not seem to have a big effect on the file size—until all unrelated nodes are removed from the file. It is logical that the osmosis --un switch needs to traverse to the end of the file, before it can release nodes, and it works much better to add this step at the end to deflate the file significantly.

At the end of the country batch, Germany has been split into separate files for

```
networks: highways railways waterways
   carto: forest farm residential industrial wood water
   admin: cities borders
 rest of: Germany natural landuse
```

Map statistics. From the map algebra, using the file sizes, one can easily detect overlaps. When extracting layers for the first time, the map engineer is curious to know what is actually left in the `*.rest.osm` files. To find this out, another switch `-stats` is hacked to the `OSMparser`.[3] The output file is directed to a `dummy.osm` file, while the statistics are written to the command line to be directed to a log file: quick and dirty, yet spontaneous and useful. With the following command line, the statistics will be logged to a file (see `statistics2010.log`):

```
../products/GER>statistics statistics.log
```

Statistical tools can only indicate error sources, yet they are very useful for regression analysis and attributes deviating more than 5% from a prior production can indicate a problem. To avoid getting lost in the details, the reader should make his own analysis by copying the statistics to a spreadsheet and ordering them by occurrences. The `clean.bat` script should give an idea on how to shrink the file sizes step by step. For reasonable sizes, the script runs quickly and can be configured to remove all temporary files at the end. In the process of creating cleaning strategies for dedicated files, the map engineer can collect unnecessary attributes for resulting products and possibly move the cleaning up the hierarchy.

Going further. Naturally, there are endless ways to compile a country, and the batch file might provide some inspiration. Yet, we are still quite some way from a final PSF. Osmosis has six relevant switches -nk -nkv -wk -wkv -tf -un to process OSM files, while the OSMparser has the "regEx" window to get the desired output.

Using the layer files, you should experiment with regular expressions to do the following:

- reduce all city tags to `place`, `name`, `is_in`, and `population`;

- reduce the cities to those with `population` > 500,000;

---

[3]The implementation will become transparent in Section 8.3.2. Note that the private `OSMparser$Occurrences.class` has to be copied with the `OSMparser.class`.

- reduce highway file to leave only `motorway.highway` tags;

- reduce railway file . . . .

The goal is to balance the reduction of map data with, for example, the rendering in Kosmos to get a smooth performance (and space) distribution. Since the aim of the `GER` project is a full country map, all details on the city level can be reduced by `building` related attributes, etc.

In the product-definition phase, an interactive tool like JOSM can also help a lot: The 30 MB file `germany.highways.osm` can be opened with JOSM, <CTRL><A> selects *all* the map data and all attributes are shown. By clicking an attribute line and <DEL> the attributes are removed from the visible map. On the other hand, this takes longer than parsing it and the resulting file is actually growing, so it should be `-clean`ed afterward.

At this point of the book, we no longer will be processing really big files. Now, the product designers can interactively process the smaller files and combine them as needed. With only a few minutes of processing time, an open file explorer, command line, and text editor, this phase can become addictive.

Using the material covered in the preceding sections, the reader can build his own Germany image with Kosmos. The directories

```
../resources/OSM.compiler/products/GER/carto/pix
../resources/gps/GER
```

provide some sample pictures and the `GERcarto.bat` lists how to create an image via command line—after having set up Kosmos interactively.

While the drawing of the background (the big files) is left to the reader, the other files have been crunched, cleaned, visually checked, and manually modified to finally extract three networks with a number of cities.

```
admin      121,034    cities.osm
networks    32,316  highways.osm
            38,879  railways.osm
            27,066 waterways.osm
```

A look at the network files in JOSM reveals the simplicity of the networks. In the process of prototyping, it is helpful to have a look at some nodes. For a real-world application, it would be worth the work to stick to the original geometry of the networks and inject an additional tag

```
<tag k="roaf" v="Navigator"/>
```

to the online OSM map to provide a public `roaf` map. Then, the network would closely match the real world, and the real objects could be localized on any map, and the compilation process could be greatly simplified.

8.3 Networking

The extraction of a *navigable network* from a map delivery restricts the output to *navigable line features*. The number and complexity of links depends on the type of navigation system, type of vehicle, etc. The number of drivable roads is much larger for passenger cars than for large and heavy trucks; boats can only use waterways; trains have to follow railway tracks.

Regardless of the network type (road, water, rail, etc.), a solid network should be *closed*. A network is *closed*, if any point on the network can be connected (routed) to any other point. The quality of the network depends greatly on the vendor's specification. In OSM maps, the *road classification* is provided as follows:

```
<tag k='highway' v='motorway'  />
<tag k='highway' v='primary'   />
<tag k='highway' v='secondary' />
<tag k='highway' v='tertiary'  />
```

For the system, a neighborhood is a network just like the network of major highways. It can apply the route calculation to the neighborhood (tertiary) to find the next "connector" to a higher-level network (secondary). Most navigation systems apply two concurrent calculations: one from current position to destination and the other vice versa.

A hierarchy simplifies the route calculation by traversing only one level at a time. A bottom-up and a top-down connectivity of every single address depends on the *arterial* classification of the field teams. With their local knowledge, they have to decide how important each road is for routing. The classification can be based on a number of criteria like speed categories, etc. The network can be enhanced with hints resulting from off-line brute force calculations.

8.3.1 The Missing `<link>`

On page 84, we stated that spatial data is represented in terms of points, lines, and polygons (on the lowest level 0). Navigable networks only require "line features," not polygons. Geometrically, a line is the (shortest) connection between two points and OSM paths are usually composed of many lines, or in navigation jargon, *edges*. This is fine to collect data, but not sufficient for a network.

The OSM data format does not define an entity between edge and way: *the link*. In terms of map data, a *link* is a polyline (sequence of nodes) with a common set of attributes.

In practice, a field team can initially collect a road's geometry with one `<way>`. Back in the office, the way is compared to existing map data and split into links. Currently Europe is typically represented by roughly 100 million links in commercial maps.

For a network, a subset of map data filtered with line attributes, the link represents the atomic unit of navigation. In terms of network data, a *link* is the connection between two manoeuver points.

In normal jargon, a manoeuver describes any special activity on the steering wheel, like a sharp turn or passing another car. In a navigational context, a manoeuver point also refers to a point, where the driver can or has to make a decision to continue his journey: for example, at a normal intersection: turn left or right or go straight ahead.

Links are also created to assist the driver in unclear traffic situations and the length of a link can range from a few meters to a few hundred meters with hundreds of shape points. As an example, if you miss an exit on the German Autobahn, you will have to drive along the link to the next exit to make a turn manoeuver and drive back to the missed exit. This can easily cost you 20 to 30 kilometers of extra travel.

Figure 8.1 shows the elements of a digital map:

- Shape points are connected by edges to define geometry.

- Ways are composed of a number of edges.

- Links connect two manoeuver points.

- Destinations are addressable points.

The illustration shows three independent networks connected only by the destination points. Imagine a person traveling by car, train, or boat with the freedom to change the vehicle being used in cities. The route from origin A to destination D defines two links A-C and C-D via manoeuver point C.

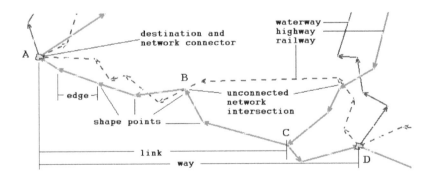

Figure 8.1. Entities of a digital map.

For development purposes, the three networks were manually merged with the cities (destinations).

> Please open the file `../products/GER/networks/germany.osm` with JOSM.

A look using JOSM reveals that the three networks are only connected on the city nodes, which have additional address data. It can also be seen that the ways are *not* split into links. In order to create a "clean" network, the ways will be chopped into links later.

8.3.2 A LinkCompiler to Assemble OSM Links

Although the file `germany.osm` contains three networks and a number of *destination* points, it is not yet suited for effective route calculations. The basic idea of a `LinkCompiler` is to go through all OSM `<way>`s of the input file and split each way into links connecting intersections. Figure 8.1 shows part of the Germany network file. By clicking on any edge between A and C and between C and D in JOSM, one can tell that the ways have not been further split into links.

From the Java perspective, the `LinkCompiler` is a great playground to get more familiar with the `Collection` framework:

THE JAVA TUTORIAL > COLLECTIONS > INTRODUCTION

> A collection   sometimes called a container   is simply an object that groups multiple elements into a single unit. Collections are used to store, retrieve, manipulate, and communicate aggregate data. Typically, they represent data items that form a natural group... A *collections framework* is a unified architecture for representing and manipulating collections.

> For a deeper understanding of the link compiler for OSM files, the reader should open the file `LinkCompiler.java` and set the command line arguments in the IDE as described in the "hands-on" box on page 92.

Program usage. The `LinkCompiler` has only one argument list with the way-key-value list of intended networks (OSM line features):

```
java LinkCompiler -rx germany.osm -wx germany.net.osm
   -wkv "railway.rail,highway.motorway,waterway.river"
```

By invoking the batch

```
../products/GER/networks> osm2net germany
```

the network file will be created and the links can be identified in JOSM.

Program design. The splitting of ways into links can be achieved using the `OSMparser`. It might be tempting to add this feature to the parser, but it could get confusing for map engineers. Map compilation teams can be separated into developers and producers to operate with a given set of well-known tools. A producer usually doesn't have time to deal with source codes and needs clear semantics for the command lines. The parser has a clear domain and should not be misused to compile links. Nevertheless, the link compiler is derived (or it could also be inherited) from the parser and has a similar architecture:

The parser `parseArguments( args )` parses the list of `key.value` pairs into

```
wayFeatures =
    { "railway.rail","highway.motorway","waterway.river" }
wayFeatures.length = 3
```

where the length defines the number of networks and is utilized to create a list of ways as sequential collection node IDs:

```
        List< List <Integer> >[]
        ways    = new ArrayList[ wayFeatures.length ];
        ways[i] = new ArrayList<List<Integer>>();
with ways[0] = railway network etc.
```

All OSM IDs are collected in

```
private Map<Integer, Occurrences>[] nodeDistribution
  = new HashMap[ wayFeatures.length ];
```

A `Map<Key,Value>` is a collection of keys without duplicates and associates (or maps) objects with keys. This structure holds for the OSM format, where each node has a unique ID with no specific position in a sequence. The `nodeDistribution` represents a lookup table to look up a value by a key. The idea is to store all of the `<way>`'s `<node>`s as keys and to count their occurrences. The class `Occurrences` is a wrapper object for a primitive `int` as a counter.

While `parseOSM()` is the central method of the parser, the link compiler only uses this method to read the ways into RAM. Note that the link compiler does not use the real-world coordinates (`lat, lon`). The idea is to create links by processing the node IDs without any knowledge of their (relative) geographic location or any other node attribute. All nodes are simply transferred to the output file and all relations are ignored and not written to the output.

Logically, *all* links are embedded in the `<way>`s. Thus, the compiler can traverse through *all* ways to extract links. The original OSM `<way>`s will *not* be written back to the output. The input stream is then closed, while the output stream is left open and waiting for the new `<way>`s, now becoming `<link>`s:

```
parseOSM()
  loadWay( element )
    identifyWay( element, keyValPattern )
      ways[feature].add( way )
```

Each network has its own distribution of nodes:

```
nodeDistribution[0]:    <nodeID,times>
                       ---------+-----
                           :  :
                          -95=2
                      21293824=1
                            77=3
                     255076555=2
                          -147=2
                           :  :
```

After `parseOSM()`, the input file has served its purpose, while the output file was only filled with the original nodes and is waiting for the new links. The method `compileLinks()` is the major method to transfer ways into links. Technically, links have the same structure as ways and can be stored in

```
List<List<Integer>>[] links
    = new ArrayList[ wayFeatures.length ]
```

The method `findNetworkConnectors()` makes use of the `Collection` algebra to determine the common nodes connecting the different networks—in this case, the destination points.

The method `isolateSharedNodes()` eliminates all keys from the `node DistributionMap` with only a single occurrence (`times == 1`). A *shared node* is a node with more than one occurrence.

The method `splitWaysInLinks()` is the core of the link compiler and loops through all ways and all wayFeatures. Each way is transferred into links from the first node to the first shared node, from this shared node to the next, and finally from the last shared node to the last node.

The method `isSharedNode( int feat, int id )` checks the node distribution, returns `true` if it is shared, *and* does the bookkeeping by removing one occurrence of the `id` or by removing the key completely from the `Map`, if it occurs only once.

Finally, `writeLinks()` writes the created links to the output file, statistics are collected, the closing `</osm>` tag is added, and the file stream is closed.

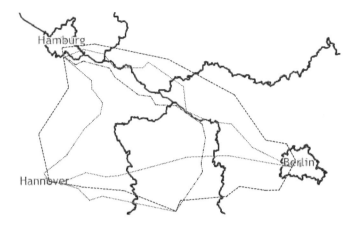

Figure 8.2. A digital map with political borders and routes between Hamburg, Berlin, and Hannover.

Note that the link compiler has been kept as simple as possible, but it is worth mentioning that for a more sophisticated compiler links should not only be split. They should also be joined if *exactly* two links share one end node.

8.4   Creating a NavigableMap

By now it should be clear that the map compilation process is very volatile as it uses various (interdependent) tools with different paradigms. While the link compiler relies on the integrity of an OSM file and is pretty numb to map data, the NavigableMap[4] will be based on the `org.w3c.dom.Document` with implied xml validations. The link compiler is part of the map compilation (preparation) step, while the navigable map will be a vital software component for navigation to proceed in the prototype development according to the vision (see Figure 8.2).

This section will focus on the *internal* development of the navigable map with the goal to create a *graph* to represent the navigable network in a software environment. The methods actually applying the graph *externally* for navigation will be added in Section 9.4

Note that the `NavigableMap.java` file is split into a top part for external methods for navigation and a bottom for the construction phase with the `main` method in the middle.

---

[4]Collections and xml structures are described in terms similar to digital map data. Do not get confused by name clashing, i.e., `java.util.NavigableMap<K,V>`!

> To get a preview of what's coming next, the main method of
> `roaf.book.navigation.NavigableMap` should be hard coded to the file
> created earlier `../resources/gps/GER/germany.net.osm` to print a sum-
> mary to `System.out`.

The `NavigableMap` has two constructors to read the network data file
or network DOM document[5] after creation. This represents the start-up
procedure, when turning on a navigation system and reading the persistent
map from an external media into the software. The duration depends on
the size of the network.

The methods

```
parseDOM( org.w3c.dom.Document osmDOM )
  parseRootNode( org.w3c.dom.Node root )
```

extract OSM XML child nodes from the file.

A node ID has the character of a look-up key, which is the structure of a
sequentially geocoded file (OSM, GDF, etc.). The `<node>`s at the beginning
of the file actually hold the geographic coordinates, while the `<way>`s only
make use of the IDs as references. In Java, this mechanism is achieved by
using a `Map` collection:

```
Map<Integer, Position> nodeMap
                = new HashMap<Integer, Position>()>
```

Note that `Position` is an interface and Java collections work with object
references. So the number of position objects remains constant throughout
the `NavigableMap`'s lifetime. After all coordinates have been collected in the
`nodeMap`,

```
lookup ID (key) -> nodeMap -> retrieve coordinates (value)
                 /     \
         < nodeID , Position >
           -------+---------
                : :
         129957469=(48.3703063N/10.884524W)
            495407=(50.8050291N/9.5378884W)
              -561=(49.4437887N/8.5421011W)
              3693=(50.3010783N/8.8246156W)
                : :
```

the OSM `ways` can be composed with node IDs and can abstract the actual
map geometry. The geometry layer is important for drawing, but as it will
turn out, not really needed for routing.

---

[5]See "XML and Java in a Nutshell" on page 50.

8.4.1   Creating a Destination List

Cartography, networking, and building a destination list make up the core
of a map product. While nodes can be looked up via IDs, a destination list
can add much more complexity as it reflects the administrative hierarchy
of a map area. If the navigation GUI can be set to different languages, then
each string has to be mapped to a geo location. If the navigation system
allows speech input, every destination entry or *grapheme* in every language
(not to speak of dialects) has to be mapped to one or more *phonemes*.[6]

For the initial `roaf` project, `extractOSMnodes( NodeList mapNodes )` creates a simple representation of a database (table):

```
List<String[]> mapGeoDB = new ArrayList<String[]>()
```

The array list is an extendable table and the first row (`index 0`) holds the
hard-coded field names:

```
final String[] dbFieldnames
  = { "nodeID", "city", "state", "country", "continent" };
mapGeoDB.add ( dbFieldnames );
```

The first column contains the unique (primary) key to look up coordinates and uses the hard-coded filters `COUNTRY, CONT` (analog to a `sql WHERE`
clause).

For a more complex destination list, OSM data references the `openGeoDB`,
which can be downloaded. As navigational software becomes prevalent in
remote devices, destination databases will exist independent of map data.
Navigation systems will be able to access databases (of any size) with local
knowledge.

In order to locate a destination, the destination entry has to be associated to a node. The process of finding geographical coordinates for an
address or vice versa is called (reverse) *geocoding* or *geomatching*. Geocoding is the common practice to enrich maps by mapping address lists (stores,
hotels etc.) to the geometry of the map.

For the `GER` product, cities are represented by a single node, while commercial map vendors make sure that *every administrative area has a single
node representation* in the network. A "city center" is carefully selected for
every place. Once all named places of the OSM file have been parsed into
the array list, the list can be transferred into the two-dimensional fixed-
size array `destinations`, which is printed in the main method. Navigation
systems usually store these strings in separate, large and indexed files.

8.4.2   Building a Graph

After all OSM nodes and destinations have been parsed into their data
structures, the method `extractMapLinks( NodeList mapNodes )` retrieves the

---

[6]Search for SAMPA (Speech Assessment Methods Phonetic Alphabet) to learn more.

OSM `<way>`s (hopefully links) by their network feature

```
int feature = setFeature( way.getElementsByTagName("tag"));
```

with the feature being the index to *one* of the *hard-coded* values:

```
private final String[] linkFeatures
  = { "highway.motorway", "railway.rail" ,"waterway.river" }
```

Then the way IDs are collected from the file,

```
List<Integer> nodeIDs
        = getWaypoints( way.getElementsByTagName("nd"));
```

node's (`GeoPoint`s) are looked up and stored in another list,[7]

```
List<Position> nodePosition = getNodeList( nodeIDs );
```

and all `from` and `to` node IDs (both ends of a link) are stored in a set:

```
Set<Integer> linkNodes = new HashSet<Integer>();
```

The collected information is used to create a `Link` as an ordered list of `Position`s and a feature:

```
Link link = new Link( feature, nodePosition );
```

After the link has been created, *only* the from and to nodes are saved, while the intermediate points strip off their IDs and loose their addressability by ID! Finally, all links of the file are collected in another set:

```
Set<Link> osmLinks = new HashSet<Link>();
```

After all links have been collected, the `nodeMap` is significantly reduced to the outer nodeIDs of the links:

```
Set<Integer> nodes = nodeMap.keySet();
nodes.retainAll( linkNodes );
```

to hold all addressable nodes reachable via destination and ID or via geo-coordinates (spatial index) to node.

The whole idea of a clean link collection is to restrict the map to its maneuver points. These are the only points where the user (or driver) can actually make a choice to navigate to one of the connected maneuver points. Building a *graph* is the way to model the connectivity. There are endless ways to build a ((non)directed, (non)planar) graph. The idea is always the same: Each node should be accessible as a `from` node of a routing algorithm. Once the node has been identified it can be looked up in the

---

[7]Note that Collection variables can be easily printed by adding a simple `println` statement for a deeper understanding.

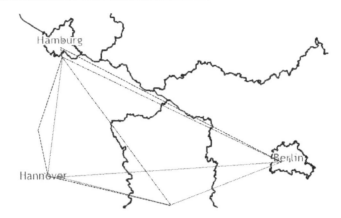

Figure 8.3. The graph of the network connections between Hamburg, Berlin and Hannover.

graph to identify the connected neighbor nodes. The links are abstracted to *edges* of the graph.

A graph is the essence of a (road) network and the heart of a navigation system (see Figure 8.3). If the network is not closed a system could run hot trying to find a way out of a neighborhood. If bridges get lost during the compilation process the transportation network is reduced to the borders of ancient times, when rivers marked borders.

`createMapGraph()`. The graph for the `NavigableMap` is created with the method `createMapGraph()`.

With the gathered information, every `Position` of the map can be looked up via the `nodeID` in the `nodeMap`. Nevertheless, the collection has to somehow browse over a number of nodes to identify one, while an array is organized in *ordinal numbers* to access the elements directly. For the sake of performance and handling, the IDs are reorganized in a sequential order. A map that associates old, random to new, sequential IDs

```
Map<oldID,newID> newNodeIDs = new HashMap<Int. ,Int.>();
```

is used to update the destination list's IDs and to create a new unmodifiable `nodeMap` with sequential keys.

Next, one empty set of `Links` is created for each nodeID:

```
Set<Link>[] nodeToLinks = new HashSet[newNodeIDs.size()];
for (int nd = 0; nd < nodeToLinks.length; nd++)
    nodeToLinks[nd] = new HashSet<Link>();
```

Each set is filled with each link connecting to the node; from and to are relative and exchangeable in this context and the nodeIDs are updated to sequential IDs on the fly. The `nodeToLinks[nd]` array returns a set of links for every node `nd` without a lookup operation, and the first dimension of the graph is complete. The actual two-dimensional graph is created with

```
mapGraph = new Link[nodeToLinks.length][];
```

The second dimension of the graph has *not* been defined yet. It could be created once using the maximum number of links for a node to leave a lot of `null` elements in the array. Here, the special implementation of more-dimensional arrays in Java comes in handy. Internally, Java only works with one-dimensional arrays and additional dimensions are modeled with arrays of arrays. This can be used to create individual arrays for every `from` node (first dimension):

```
mapGraph = new Link[nodeToLinks.length][];
for (int from = 0; from < nodeToLinks.length; from++)
    mapGraph[from] = nodeToLinks[from].toArray( new Link[0] );
```

The result is a two-dimensional array `mapGraph` of links:

```
Link[all nodes in sequential order][reachable nodes]
```

The internal development (private methods) of the navigable map provides the platonic structure for navigational operations:

1. external selection of a destination D in `nodeMap` via list or via geo coordinates;

2. finding the number of neighbors via `mapGraph[D].length`;

3. evaluating each connecting link (distance, feature, etc.) via `link = mapGraph[D][neighbor]`

4. traversing the graph until one of the neighbors is identified as the destination.

Finally, the `Link` itself describes *how* to drive from a node to a neighbor. The private visibility of the graph prevents direct access by external clients. (The next chapter will introduce customized basic navigational methods.)

8.5   Conclusion

The basics of map compilations have now been outlined and the reader should be able to configure his own map compiler to produce a PSF for his own application. The work of a map engineer, however, is dominated by planing. Any fix should be applied before going home in the evening in order to run a test overnight and get the result in the morning. The worst case scenario is kicking-off a compilation on Friday afternoon to use the machine power on the weekend and to find on Monday morning that the fix was not checked back into the versioning system.

Setting up the first cycle of a compilation consumes machine and man power and consecutive cycles or productions have to be observed carefully. How independent are the branches of the compiler? How vital is the missing information for the rest of the compilation? Where can the compiler be re-started after a failure? Is there a map or a compiler problem? Is enough disk space available?

Compiler Paradigms

As map data is constantly growing, sequential formats become more problematic to handle. The files have to be unpacked, processed, repacked, and stored. UTF coding has to be used to represent international languages.

The `OSMcompiler` was use to save the reader from installing a relational database management system (RDBMS). Nevertheless, it must be mentioned that the paradigm to process sequential files has given way to relational data manipulation. The advantages are obviously the guaranteed referential integrity of the delivered map data. After the data is loaded into a relational database, the constraints can be enabled to validate the integrity.

Once map data is in place, every compilation step can be modeled with SQL scripts to produce intermediate tables (data states). Every conversion step can be implicitly controlled by the database system. In addition, most RDBMS have a built-in engine to process spatial data to calculate distances, areas, and to simplify shapes, etc.

Therefore, it is strongly recommended to build any map compiler in conjunction with a RDBMS. Still, post-processing is inevitable, since SQL is not suited for route calculations, etc. Basically SQL performs best by filtering data records, which are represented in table rows (horizontal). To compose a route of links, the database has to vertically connect a number of rows (of links) to a path.

PSF and the Navigational Data Standard (NDS)

Whatever compiler paradigm is used, in the end the data has to be exported to a PSF with a well-defined file structure on the media. Historically PSFs are static data and map compilation is always a one-way process. Every navigation system for different cars with different features and different coverage has its own PSF, which is created by large map compilation teams.

Since the production of a new system requires many resources the German car industry has created a committee to define a standard PSF, the *navigational data standard* NDS. The NDS is defined by a number of XML files to be loaded in different systems, which have an internal DB engine in the target system.

The Dream of the Incremental Compiler

It is apparent that every new map delivery has to go through a full compilation to reveal data changes and problems. This is not very satisfactory when you know that digital map releases only deviate by a maximum of 10% per quarter. This raises the idea of an incremental map compiler, which can identify the changes in a map and patch them into the previous map product. Besides saving compilation time to reduce the time to market, an incremental map update on the target opens a wide range of new business models for dynamic map updates.

Nevertheless, a map compilation is a dedicated process with a tool chain of different manipulation programs. Also, a map vendor can not guarantee identical ID spaces for links and other map features and this makes it very hard to identify map changes.

For your own projects, you might consider extracting a stable network from the map and run a delta process on the latest map release. The network can be optimized to the target application, enriched with navigation hints, etc.

Chapter 9

Navigating Objects

9.1   Introduction

Although moving objects reflect a realistic trace on the map, they have
no freedom to deviate from the prerecorded or live data. By extracting a
route from a digital-map network, the navigating object can make a choice
at the maneuver points.

A clean network with a related administrative database enables a user to
extract a well-defined route from the network. Again, we can close the door
to all of the theory of networking digital maps and finally make use of the
products created on a higher-level to calculate routes. Route-calculation
algorithms assist the user to navigate through a network by evaluating the
best possible route. A `GameMap` is developed to traverse the graph of the
`NavigableMap` produced in the last chapter.

Current navigation systems are dedicated computers used to collect all
data relevant for a route calculation according to parameters and external conditions. To get an idea of dynamic route calculation, a `Navigator`
will visualize a number of objects navigating between the larger cities of
Germany.

9.2   Navigation Systems

One of the first built in navigation systems was introduced in 1996, when
digital maps covered only a few major cities and their connections. The
navigation systems operated completely off line, and the map data (PSF)
was provided on external media. Today, most industrialized countries are
nearly one-hundred percent geocoded, the devices are scaled down to *personal navigation devices* (PNDs) and are moving to smart phones with the
ability to deal with dynamic integration of additional online content to
alter a route.

The amount of constantly growing map data and the restricted memory of handheld devices suggest distributed functionality between on- and
off-board calculations on a server. The value-add will be *location-based*

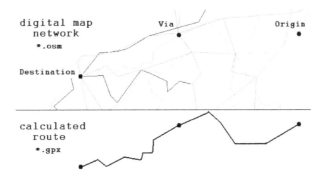

Figure 9.1. From the project view a navigation system can read a digital map and calculate (extract) a route for objects to drive along.

services, like *today's* menu in a nearby restaurant. The users can navigate directly to a convenient location and book a table. The integration of internet information and services will finally close the gap.

For the ROAF project, the `NavigableMap` represents the network, which can be evaluated with the `GameMap` to extract a route according to the real object's input. The real object is actually able to dynamically navigate through the network by providing destinations.

In the long run, the architecture of the `RealObject` provides the freedom to implement *any* navigation system to extract a route to drive on. Mathematically speaking, the navigation system operates on the network to return a route (see Figure 9.1). Also, the digital map source is *not* predefined in any way. The only constraint is that the digital map data should realistically reflect the real-world geometry. By routing two ROs with different navigation systems and different digital map sources through a ROApp, the real objects are actually evaluating their sources against each other.

For a more complex navigation system, the reader should have a look at the OSM project, `Traveling_salesman`:

`wiki.openstreetmap.org/wiki/Traveling_salesman`

    Traveling_salesman is a routing and navigation program for the OpenStreetMap.

    It is written to encourage experimentation.

    Nearly every part of it is developed as a plugin and can be replaced:

- route-calculations (multiple easy to understand algorithms already supplied),

- metric to optimize a route for (e.g., "most fuel efficient route"),
- finding places,
- ...

Alternatively, one can build one's own system by using the map compiler and adding complexity step by step.

9.3 Route Calculation

The Origin

The initial task of a navigation system is to determine its current position. Where am I with respect to the current route? What maneuver is next?

Today *vehicle positioning* has become much less complex, since the GPS signals are very precise and digital maps supply a high density and coverage. Once the vehicle has determined its coordinates, the position can be snapped to the road network. The closest map node relative to the current position represents the `origin`, while a `destination` is provided by the user/driver. In the `roaf` project, vehicle positioning does not really take place, since the objects do not rely on satellite signals or any sensors. The "signal" of the `GPSunit` is always available and always precise. Position and heading are known at any instance in time.

The Route: `origin` – (`via[]`) – `destination`

With the origin and the destination, the system can begin to search for a route on the given road network. A number of additional *via points* can be formally treated as a number of routes combined on a higher level in the system. The core process of a navigation system is the traversal of a graph representing the network. While the graph is static, routing is dynamic and re-routing can occur whenever the vehicle leaves the precalculated way. Route calculation is based on a graph composed of nodes and edges. Neighboring nodes or *vertices* are *adjacent* to each other. The number of neighbors grows quickly and requires a search strategy to exclude nodes as early as possible. This can be achieved by defining a *cost* for every connection.

The most widespread algorithm to search a graph is *Dijkstra's shortest path algorithm*, which is used in navigation systems and in other applications, such as games. The *A-star algorithm* is a special implementation of Dijkstra and makes additional use of the destination's coordinate. The proprietary implementation and best-kept secret of each system is the black box for the cost calculation. A full-fledged navigation system can combine computations about time and distance, consider time of day, day of week, speed limit, restrictions, and real-time traffic information (TMC, traffic message channel, etc.).

9.4 Exploring the Graph with a `GameMap`

For a basic understanding of (or as a starting point for) routing algorithms, the next practical steps will be implemented in a `GameMap`. The idea is to use the `NavigableMap` to set up the graph and then use the `GameMap` to traverse the graph *without* considering heuristic costs. On a *game board*, each *edge* can be related to *one move*, regardless of its distance. Mathematical formulas are avoided in favor of discrete node IDs without location information. The cost, considering vehicle type, dynamic barriers, time and speed dependency, preprocessed hints, etc. could be added later by consulting the `NavigableMap`.

> Again, we suggest that you run the main method of `roaf.book.navigation.GameMap` hard-coded to `../resources/gps/GER /germany.net.osm` for a preview.

Class Design

The `GameMap` extends `NavigableMap` and overrides the constructor, which creates a graph from an OSM file (i.e., `germany.net.osm`) as described in Section 8.4.2. Since the creation of the graph is private to the `NavigableMap`, the reader can focus on its `protected` and `public` members and methods.

`getNodeMap()`. First, the main method retrieves the node map via `navMap` `.getNodeMap()` (see page 115), which supplies all addressable origin and destination nodes of the network. In order to find the way from node O to node D, two search strategies can be used: search by depth or search by breadth. Most navigation systems implement *shortest path algorithms* by searching many destinations from the origin and vice versa. The search results are combined in a distance matrix for all vertices (nodes) of the graph. Shortest times can be derived by dividing the lengths of links by speed (map feature).

All links have indexes of the network and their names and number can be retrieved via the string array `linkFeatures`.

`getAllNeighbors( origin )` and `getNeighbors( origin, feature )`. These two methods can be seen as platonic routing functions that are implicitly used by higher-level algorithms. From the programmer's perspective, these methods abstract the direct access to a (special type of) graph and restrict the routing to node IDs. Both methods return a `Set` of `Integers`. The methods are private, since the results can also be retrieved with higher-level methods.

`getDestinations( origin, nrMoves )`. The method `getDestinations` is the higher-level method to determine all reachable nodes with a given number of moves (edges) without duplicates. The usage `getDestinations ( nodeID, 1 )` is analogous to `getNeighbors( nodeID )`.

Note that the method `getDestinations()` of the navigable map returns the destination "database." The destination list with nodeIDs is provided as a two-dimensional string array with five columns, the first row holding the column headings and the first column (starting with the second row) holding the integer node IDs.

`getMinimumMoves( origin, destination )`. Before finding *a* route from O to D, it is helpful to find the minimum number of edges (moves) between them. This can be achieved by exploring the graph over the full *breadth* of each node's neighbors starting from one origin and ending at one destination. Both game-map methods `getDestinations` and `getMinimumMoves` have the same structure with different return values. The pseudocode looks like this:

```
1. collect all neighbors of the origin (from node)
2. check if any neighbor matches the destination
   if destination is not found:
       3. move all neighborNodes into the fromNodes and
       4. remove all visitedNodes, then go to point 2.
   if destination is found:
       5. return the minimum number of edges
```

Note that the minimum number of edges does *not* necessarily reflect the minimum distance of the shortest route (as shown in Figures 8.2 and 8.3)!

`getPaths( Integer origin, Integer destination )`. This method implicitly applies `getMinimumMoves` (which implicitly applies the method `getAllNeighbors`) as the route length (not distance!) to determine one or more paths with exactly this number of moves.

The implementation introduces another collection, `java.util.Queue` a collection designed for holding elements prior to processing.

The main reason for this construct is that elements can be removed without specifying an object or index (order is irrelevant in this context). The idea is to fill one queue for each move from the previous node and loop from the origin over all moves and retrieve one node for every move. As soon as a queue is empty, the next one is filled until they are all used up.

Sometimes a look at the data (structures) is more helpful than the pseudocode. The graph is transformed into a tree view from the origin's perspective:

```
graph -> tree view  ->  collection view  ->  move loop
                                         currPath[] queues[]
    origin: 1                move#: 1  2  3    [0] = origin
move#     / | \              :   1  2  5 11    [1] = [0]
   1    2  3  4     currPath  1  2  6  X    :
        /| / \ |\            :   1  3  7 12    [2] = [1]
   2   5 6 7 8 9 10          :   :  :  :  :    :
       | / \ |  | \          :   1  4 10 16  [moves] =
   3  11 12 13 14 15 16  path [0][1][2][3]         [moves-1]
```

The `GameMap` implementation traverses all paths. For an effective navigation system, this search should be optimized (by removing visited nodes, etc.), since even the `germany.net.osm` map has more than two million paths for only nine moves.

Depending on the application, an additional feature filter could be added for each method:

```
getDestinations( origin, nrMoves,     feature )
getMinimumMoves( origin, destination, feature )
       getPaths( origin, destination, feature )
```

Finally, after the node IDs are determined, the actual route can be retrieved from the map links provided by the `NavigableMap`.

`validateMap()`. This method is private and can be used to handle quality control for a digital map. Nevertheless, this method is a very optimistic brute-force method traversing *all* connections of the graph. Note that the `GameMap.main` method includes a validation of the entire network and takes about three minutes. Of course this line can be commented out. As the map gets larger, the validation time approaches infinity. In order to validate only one entire city, large machines, search strategies, and network layers come into play.

`findNodeID( Position pos )`. This method is basically for use with a visual map. Since it is practically impossible to click on the precise `lat` and `lon` values of a node, there has to be a mechanism to find the closest node to a given point. Usually this requires yet another collection of all the nodes, where each node can be looked up with a *spatial index*. For the time being (and for a small number of nodes), the method implements a simple, although slow, loop over all nodes.[1]

---

[1] Check out the java spatial index, jsi, at `sourceforge.net/projects/jsi`.

9.4.1 Conclusion

The `GameMap` methods for breadth and for length demonstrate the very basics of routing and should enable you to use more sophisticated algorithms to traverse a graph. Most implementations vary single source to single destination, single source to multiple destinations, and multiple source to single destination or a bidirectional combination of algorithms to meet somewhere in the middle. Of course, navigation systems have to add the real distance (cost) instead of finding a number of edges. (There is infinite cost for the wrong way of a one-way street, etc.)

The A star algorithm adds the Euclidian distance from origin to destination to take the deviation into account. The algorithm can be improved step by step to add *pruning* to remove branches as early as possible and identify the better ones by some heuristic information. The routing algorithm can also be supported by the map specification in the form of map layers or aggregated segments (composite feature). In the end, no routing algorithm is perfect and each implementation has to be fine-tuned (performance vs. quality) and configured by the end user (avoid highways, etc.) In most systems, the routing algorithm gets a certain slice of the CPU to allow a maximum amount of time, before providing a route. The user of a navigation system actually "feels" the time from selecting a destination until he can start driving and the time for re-routing, when leaving the precomputed route.

All implemented methods start from scratch, while most navigation systems have a routing thread to make better use of the provided CPU power. A tree is built from the current origin and every branch is complemented as far as possible. Every time a move is made the tree's origin is moved up and alternative moves are dropped. The implementation of an extra search thread has the advantage that the tree is not lost after a decision which saves time.

The processing of digital maps opens the world to the real object. Starting out with road maps reduces the globe's surface (500 million km^2) by water (70%) and deserts (30% of land area) to the "car countries." Even if the real object should not represent a vehicle, the world's road network can be perceived as the modern map grid. Of course, the real-object architecture is fit for a map of airplane traffic as well or to process ship data (search for AIS: Automatic Identification System).

9.5 The Navigator

The `Navigator` sums up the compilation of navigable maps and uses route calculation on `NavigatedObjects`. While the `MovingObjects` application can be used to observe the recorded traces of a motorcycle, the `Navigator` can be used to create and observe navigating objects.

> Please go through the quick start instructions: `../deployed/Navigator.txt`. The `Navigator.jar` can be launched without a predefined area and directed to `../resources/gps/GER/` while the files are hard-coded in `Navigator.java` with the variables: `path gpxFile smallFile bigFile`.

Both applications are basically designed in the same way, and the reader should be able to identify the major application classes developed in this chapter. A closer look at the `NavigatedObject` reveals that it is actually *not* a navigat*ing* object. While the motorcycle `Buell_XB12Ss` is using `replayGPStrace` to replay a recorded trace, the `NavigatedObject` is replaying a calculated route with calculated speeds. It is actually not aware of a digital map and relies on the routes provided by a navigation system from a navigable map.

Nevertheless, the external observer can not tell the difference and the next chapter will move the navigation system into the real object. A player object will be able to use the game map to apply a strategy for dynamic and intelligent routing—internally.

Part IV

ROApps: RealObject Applications

The abstract `RealObject` serves as the programming shell to implement the behavior of any physical object. A simulation of a physical object has to adhere to classical physical laws. These laws have to implemented inside the `RealObject` and can be validated by observing the object from the outside. Vehicle traces can be compared to digital maps as a well-defined environment to make sure the trace is plausible.

In order to decouple the real-object development from its environment, it has to be separated from the application logic. Objects and scenarios should be developed independently of each other, bound only by real-world constraints.

A *real-object application* or ROApp is a Java server application composed of connected `RealObjects`, just as Java applications are composed of various `Objects`. A RO can log on a ROApp to participate in the scenario and interact with other ROs.

After the definition and (minimum) implementation of the ROApp architecture, the RO and ROApp development will be put together into one distributed application. The reader can launch a `GUIPlayer` to participate in a game running on a separate Java Virtual Machine (on a different computer). The programmer can implement additional players using dynamic routing to play against other users.

Chapter 10

Separating the RO Client and the ROApp Server

10.1   Introduction

As an application grows to a certain critical size, it becomes cumbersome to remember all details in various contexts. A good development team is characterized by a good distribution of know-how and responsibility. The same principle holds for good software. A distributed application can help to decouple semantic modules.

Using Java remote method invocation (RMI), the real objects will be engineered into *client* programs, which can be launched independently on one machine and connect to various real-world *server* scenarios in the network. The `roaf.book.rmi` package is a minimum implementation to run `MovingObjects` on two JVMs. It serves to develop a terminology in the context of ROs and ROApps and abstract technical details.

10.2   Observing Remote Objects

In Part III, we introduced two simple applications to observe real objects. The `MovingObjects` and `Navigator` applications basically share the same design. In one scenario, the ROs represent motorcycles and replay a recorded trace, while the navigable objects have a choice of paths. The external observer can not really discern a difference.

Both applications can be logically subdivided into three parts, or `Thread`s:

1. `MapViewer` for visual observation → Swing `Thread`;

2. `Sampler` as a receiver of the object's GPS data → server `Thread`;

3. many `movingObjects` of type `RealObject` → client `Thread`.

Threads can be seen as a technique to simplify the separation of a program into independent parts. Each thread represents a different—virtual—CPU. Nevertheless, all threads run on the JVM, where they are launched. *Distributed programming* is the next step after threading and Java supports distribution via *remote method invocation*:

> THE JAVA TUTORIAL > RMI > INTRODUCTION
>
> The Java Remote Method Invocation (RMI) system allows an object running in one Java virtual machine to invoke methods on an object running in another Java virtual machine. RMI provides for remote communication between programs written in the Java programming language.

Our vision has anticipated this separation:

1. *RO vision* (page 18): Technically a `RealObject` should be an independent program to be run independently on its own JVM (and CPU).

2. *ROApp vision* (page 19): Technically a RO is a client and the ROApp is a server application.

3. *mapping GUI requirements* (page 56) . . . the GUI should not interfere with the simulation and should only observe—as long as CPU power is available.

> To get started, please go through the quick start instructions: `../deployed/RMIROApp.txt`. Note that `RealObjectClient.main` is hard-coded to a GPX file, which has to be adapted to a new environment.

10.2.1 ROApp Server Design

roaf.book.rmi

The classes of the `roaf.book.rmi` package comprise a minimum implementation of a distributed Java RMI application—in addition to the familiar components of `MovingObjects`. The first step to turn a class (or application) into a server is the following:

```
class RealObjectsApplication extends UnicastRemoteObject
```

Technically, the `UnicastRemoteObject` turns the class into a Java RMI server as provided by the package `java.rmi.server`. Although the constructor is basically the same as the `MovingObjects` constructor, its super-constructor implicitly creates and exports the object to the RMI runtime and turns it into a server object to be referenced via network connections.

The `main` method creates a `Registry` to represent the actual network address, where remote objects can be "published" or *exported* for interaction. For the demo application (and simplification) on a single computer, a `SecurityManager` is not used. Therefore, policy files are *not* required at this point to avoid supplying the absolute path to the code base via command line.

After construction of `RealObjectsApplication`, the server is online. Before clients can look up a *reference* to a remote-server object, it has to be bound to a registry to be looked up with the registered string (with host and port). Note that the (un)bind button only toggles the reachability for clients. The clients who are already connected stay connected.

RMI hides the network protocols from the programmer and abstracts them to the well known signature (API) of any other Java object. The process of defining the object communication is the same as before: define the semantics via interfaces and implement them:

```
RealObjectsApplication ...
        implements RemoteServer, RemoteObjectsApplication
```

This class demonstrates the implementation of two interfaces for different semantics. The `RemoteServer` represents the technical part of the communication to obtain a connection from client to server and enable the client to retrieve general information like the server's time and space implementation.

With the remote interfaces, the initial Java vision as a remote-control language (page 4) is complete. A Java object can obtain a reference to a server, e.g., a television set, and invoke commands on the `RemoteServer` (or stub in Java terminology).

Besides passing stubs, RMI also allows the transmission of Java objects. Since these objects are transmitted *by value*, the programmer no longer has to worry about unwanted references; the object is copied from one JVM to another. The programmer can approve any object as an "exchange particle" by tagging it as `Serializable`.

With the `RemoteServer` signature,

```
GPSinfo getGPSinfo() throws RemoteException
```

the connected client can receive a copy of a `GPSinfo`-type object without knowing the actual implementation (in its `import` statements) and can apply its `distance` metrics.

The prefixes `Real...` and `Remote...` of the classes reveal their role in the ROAF:

```
RealObjectsApplication implements RemoteObjectsApplication
```

The *real*-object application is running on one JVM, while the *real*-object clients communicate with the *remote*-object application. Vice versa, the application communicates with *remote*-object clients. From a programming perspective, the client-server architecture is blurred and is basically given as an $n : 1$ relationship for the RO:ROApp pair.

10.2.2   RO Client Design

The RO is wrapped in a client program responsible for RMI networking:

```
RealObjectClient extends RealObject
```

The client is started on a separate JVM with its `main` method doing the handshake with the server step by step.

The server reference can be looked up and cast to the remote-server interface:[1]

```
RemoteServer roaServer =
        (RemoteServer) Naming.lookup("//my-PC:1098/ROA");
```

Since the interface is part of the client and server software, it can be used to store common `SERVERNAME` and `PORTNUMBER`. Once the client has connected to the server, it can retrieve the server's `GPSinfo()` implementation.

Retrieving space and time from the server is a unidirectional request initiated by anonymous clients. The server can only use the return values to respond. As the vision requires the RO to report its position at any point in time (page 18), the server is also required to get a remote reference (a stub) to the client:

```
interface RemoteObjectClient extends Remote
```

This interface defines the `RealObjectClient` as a `RealObject` implementation. Instead of extending the `UnicastRemoteObject` to export the entire client, a private, inner class `RemoteClient` is used to satisfy the defined methods.

This way the developer can "run" a RO on a local machine and explicitly create and export the `RemoteObject` (if and) as needed. Sometimes inner classes are referred to as windows to their outer classes. The `RemoteClient` propagates the outer classes' qualities and can be transferred to the server. With the remote-servers method,

```
boolean identify( RemoteObject client ) throws RemoteException;
```

the client introduces itself to the server. Now, the server can use the client's remote methods and decide to authorize it (return `true`) to participate in a scenario by logging in:

---

[1] "my-PC" should be replaced with the server's name or with `"localhost"`, if client and server run on the same machine.

```
RemoteObjectsApplication login ( RemoteObject client )
                                throws RemoteException;
```

Now, the method returns the `RemoteObjectsApplication` interface defining the actual interaction with the scenario. Note that the interface could also be retrieved by casting the `RemoteServer`. Nevertheless, the client should apply the formal identification and login procedure for an explicit server authorization.

The two interfaces indicate a handshake in two phases. First, the client looks up the remote server to get time and space and identifies itself, and then the server adds its reference to the ROApp scenario. Note that the client's stub is not bound (published) to a registry, since it can be passed to the server programmatically. Connecting via lookup on a registry is only needed once. Finally, the client and server can inspect each other asynchronously.

10.2.3   RO–ROApp Sequence Diagram

The sequence diagram below provides a good overview of the client-server dialog:

```
server machine          any machine
RealObjectsApplication  Registry
----------------------  --------

1. getRegistry        <-- host + port      client machine
                                           RealObjectClient
2. new ROApp --> export                     ----------------

3. bindRemoteObject   --> SERVERNAME <--  4. Naming.lookup
                      |                    |
                Registry no longer needed  |
                                           |
                 <--- reference ---  5. (RemoteServer)

   new GPSinfo       ----   copy    -->  6.  getGPSinfo()

                                    7. new ROClient(GPSinfo)

                                    8. export new RemoteClient
                                           |
(RemoteObjectClient) --- reference --->  login ( client )
          <-- reference --  (RemoteObjectsApplication)

   movingObjects.add( client )              replayGPStrace

                 <--- interact --->
```

Here are the steps of the client-server dialog written out:

1. The ROApp statically (creates and) locates a registry (and port) on the host.

2. A ROApp instance is created and exported into a remote object.

3. The ROApp is bound to the registry, which makes it reachable via URL.

4. The RO connects to the registry and looks up the ROApp via URL. At this point, the client has a direct connection (reference) to the server.

5. The RO can cast the ROApp to any remote interface to invoke remote methods.

6. The RO needs to get the server space and time in order to: instantiate RO with server space and time.

7. The RO creates a remote client and passes its reference to the server via login. At this point the server has a direct connection (reference) to the client and server and client can interact asynchronously.

10.3   ROAF Client Software

With the separation of client and server to different JVM, the `RealObject` vision is technically defined. This might appear a little surprising at this point, since the RO can only provide its latitude and longitude on request. On the other hand, the mission was to come up with a concrete design—our minimal implementation is an idealized object.

Technically speaking, all clients in the ROApp framework serve the task of enriching any real object connected via `RealObject` with additional information.

The client software consists of three major components (see Figure 10.1):

1. A GPS unit to manage (UTC) time and (WGS84) space coordinates and to convert geographic coordinates to metric units (meters). The GPS software can be driven by a real GPS device or a software simulation.

2. A `RealObject` to connect hardware or to implement software to reflect real-world behavior.

3. A `RemoteObject` interface as the representative of a `RealObject` on a ROApp server. The remote server can inquire the RO coordinates at any time.

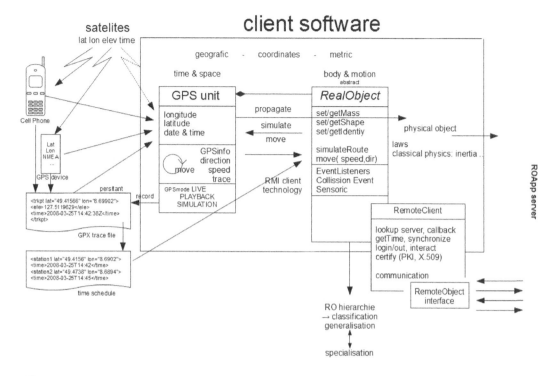

Figure 10.1. The basic components of every `RealObject` client in the ROAF.

Is this implementation fit for *any* physical object and to transmit *any* information? The answer can be found in object-oriented software constructs. By inheriting the `RealObject` and the `RemoteObject`, anything can be implemented on the client and server side. So much for now: inheritance will play a decisive role in defining the client-server architecture in the next chapter.

10.4 The `roaf.util.RMI` Class

The `roaf.book.rmi` package does not make use of the full extent of RMI applications. Therefore, the `roaf.util.RMI` class is created as a common basis for all ROAF developers to avoid redundant code and to be able to improve the RMI coding in one place. The class is designed to be final and unextendable and to provide the basic methods needed for a distributed RMI application. The static methods are designed to catch all (cascaded) exceptions, provide hints, and return `true` on success and `false` on failure. Thus, the programmer only needs to import `java.rmi.Remote` to tag classes for remote access and `java.rmi.RemoteException` to detect connection problems.

Java RMI is divided into three packages comprising the three parties involved (see the sequence diagram on page 137):

1. `java.rmi`: This package provides everything for the RMI client:

   - `Remote` interface to reference remote objects;
   - `RemoteException`, if something goes wrong in the network;
   - `Naming` class to get the registry and look up remote objects;
   - `RMISecurityManager` to identify the involved classes;
   - a number of `Exceptions`.

2. `java.rmi.registry`: This package sets up and locates a registry—the third party between client and server. The registry is only initially needed to look up the reference of a RMI object. Once the connection is established the registry could be turned off.

3. `java.rmi.server`: This package provides classes and interfaces for supporting the RMI server side. This is basically the

   - `UnicastRemoteObject`, which inherits from
   - `RemoteServer` (`getClientHost()`), which inherits from
   - `RemoteObject`

   and is able to turn any object tagged with `Remote` into a remote object.

Using the `roaf.util.RMI` class, the ROAF developer can invoke RMI methods with a single call and proceed in the program flow by checking the return value indicating success or failure.

1. `setRMISecurityManager()`: The `RMISecurityManager` should be set as soon as the application is distributed via the internet. It can only be set before beginning with RMI and cannot be unset. The consecutive methods provide hints on how to satisfy the security manager.

2. `allocateRegistry( String host, int port )`: With this method, a registry can be launched on any host and specified port (port 1099 is the default port for the RMI protocol) with the command line tool:

   ```
   %java.home%\bin\rmiregistry.exe <port>
   ```

   Then, this registry can be allocated from any other (connected) host.

   Note that the `RMI` class hides the `rmiRegistry` as a private member! Consequently, the `roaf` application should always call this method with the same host and port. Otherwise, an initial allocation of a previous registry will be lost.

3. `createRuntimeRegistry( int port )`: Creating a registry externally from the command line is usually the best choice, since client and server can be run and killed independently. On the other hand, it is very convenient (during development) to create a registry from the Java program. The important difference is that the registry only runs as long as the initiating JVM is alive.

   The method first looks for an external registry running on the `localhost`, allocates it, or creates one for the runtime of the Java program. The programmer does not need to differentiate between the two.

4. `getHost()`: Helper method, if a rmiRegistry is set.

5. `getPort()`: Helper method, if a rmiRegistry is set.

6. `getRmiURL()`: Helper method, if a rmiRegistry is set.

7. `getLocalhostName()`: Helper method.

8. `getLocalhostAddress()`: Helper method.

9. `exportAsRemoteObject( Remote remoteImplementation )`: Using this method, all real objects only need to implement a `Remote` interface plus the methods to be invoked remotely. Thus, the programmer is not confused by extending the `UnicastRemoteObject`.

10. `unexportRemoteObject( Remote remoteImplementation )`: The method reverses the export and should be used to withdraw the remote availability and before shutting down the application in a clean way.

11. `registerRemoteObject( String name, Remote remoteObject )`: Before a remote object can be addressed on a registry, it has to be *bound* to it, or *published* under a certain name.

12. `Remote lookupRMIURL( String RMIURL )`: Finally, after the remote object is exported and registered, it can be looked up by a client. The RMI URL conveniently combines host, port, and published name of the remote object.

Chapter 11

Client Server Architecture

11.1 Introduction

With a good understanding and technical specification of the `RealObject` client, we consider the server side. The server has to be able to identify, authorize, and handle a large number of clients in an ordered manner. The server design has to be open to the inheritance of simple objects to higher-level constructs with additional attributes and methods.

Similar to the extension of the `RealObject`, the `RealObjectServer` should implement the technical prerequisites for *every* real-object application. The software design should restrict the ROApp to a lean semantic layer running on the server.

11.2 Status Report

At this point, the reader should have a notion of real objects, real-object applications, and object-oriented methodologies. Large portions of the Java Tutorial Trails have been used in the `roaf.book` packages. These packages can be deleted now when the actual project sources are moved to the `roaf` root folder.

The `roaf.gps` and `roafx.gui` packages are shown in Figures 4.1 and 5.1 (pages 52 and 60) and look similar to the `roaf.book` packages. The new `roaf` project will contain the finalized ROAF library `roaf` with a sample ROApp in `roa.ldn`.

A library is part of the JVM concept. Only with an installed Java Class Library can higher-level programming tools (like the `String` class) be used on any JVM. Java provides means to install packages as a library so that the packages do not have to be defined in a class path and become part of the JVM.

Every JVM participating in the framework should run with the `roaf` packages to provide a common basis for the real-object application framework. Although this has not been enforced in any way yet, it should be kept in mind for long-term development. The Java reflection technology enables programmers to inquire classes, which are not part of the start-up procedure and which only appear at runtime.

From here on, the development will be coded directly into the new packages. The text will basically follow the call stacks and jump from client to server according to their dialog. For the reader, this is somewhat a reverse-engineering tutorial. The Java Tutorial used to have a final trail (although it is not published anymore, the trail can still be found on the internet) to collect previous trails and put them together into one application—this can serve as a guideline for putting the ROAF together.

> THE JAVA TUTORIAL > PUTTING IT ALL TOGETHER > INTRODUCTION
>
> Now, it's time to put all that knowledge and programming skill to use. You can choose *your own project*, or you can follow the description and analysis of some of the projects that we did.
>
> A client/server application that implements the game of BINGO. This example broadcasts information via a multicast socket, builds its GUI with Swing components, uses multiple synchronous threads, and communicates with RMI.

The software in Parts II and III was developed in `roaf.book` packages related to Java Tutorial Trails. The remainder of this part rolls out the server architecture with an application layer on top of it. It is much more convenient and effective to follow the text by setting up clients and server in advance and adding `println` statements in the source, while reading.

> Please set up the application environment now as described in Section 13.2.1 to run the application as needed.

11.3 Four-layer Architecture

After separating client and server, the vision gets even more concrete. The `roaf.ros` package was designed as the origin of all real objects. Each abstract `RealObject` can be extended to specify an object representing the real object; the server is the environment for the client objects.

Figure 11.1 shows the four main components of the client-server architecture. The `RealObjectServer` can be coded against abstract objects and interfaces in the RO package. From a design perspective, it is vital to stick to the package and not to apply any methods of the extending classes.

The figure supports the architect in defining guidelines. One guideline for the real-object package was already defined to reflect human and scien-

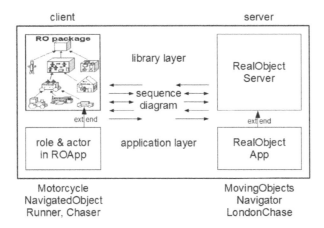

client   server

RO package

library layer

RealObject
Server

→ sequence ↔
↔ diagram

extend   extend

role & actor
in ROApp

application layer

RealObject
App

Motorcycle   MovingObjects
NavigatedObject   Navigator
Runner, Chaser   LondonChase

Figure 11.1. The four layers of a distributed real-object application: the client (left) and the server (right). The code of the common library (top) defines the ROAF laws, while the concrete application implementation (bottom) can add rules.

tific know-how—a library for real-world knowledge. The `RealObjectServer` should not be abstract, but rather should enable any ROAF developer to launch the server "out of the box." In consecutive steps or cycles, the `RealObjectApplication` extends the server and programs the scenario with its API.

The left-hand side depicts the client and the right-hand side, the server. We can also look at the architecture in terms of layers: the library layer (top) does not have to be modified by the application developer layer (bottom).

11.4 The `ServerEngine`

For the architect, the `ServerEngine` represents the external development environment on the server side. The engine is not part of the library and resides in `roa.ldn.server`, which is part of the first real-object application to be rolled out step by step. Again, the elements of the `MovingObjects` application can be recognized, although some of them are commented out. For the time being, the server engine does not create a GUI or start an observer. Prototyping is an important process to separate library and application. Code can be shifted between these two layers to generalize the library, while keeping the application lean.

11.4.1   External Configuration with Property Files

In order to avoid hard-coded relations, the server's parameters are stored
in an external properties file, which is loaded via command line. If the
properties file does not exist, it will be created. In practice, the properties
help to make the server *not* abstract by supplying mandatory parameters,
like a bounding box. The path and file must be specified as command
line arguments, when launching the `ServerEngine` along with the `policy`
file (VM argument) to set the RMI security manager—as configured in the
server batch file `LondonChase.bat`.

The class `ROServerProps` is a wrapper class for `java.util.Properties`

THE JAVA TUTORIAL > ESSENTIAL CLASSES > THE PLATFORM ENVIRONMENT

> Properties are configuration values managed as key/value
> pairs. In each pair, the key and value are both String values.
> The key identifies, and is used to retrieve, the value, much as
> a variable name is used to retrieve the variable's value.

This class extends a `java.util.Hashtable` to hold properties as pairs of keys
and values and has the ability to keep them persistent in the file system.
The class is initially designed to hold the parameters for the server's `host`
`port name` values. Once in place, the designer can easily add values during the modeling process. The class `ROServerProps` resides in the `common`
package to be extended by server, application, and client:

        `ROAppProps extends ROServerProps`

Note that interfaces can also be used to store common information, like
server name, port, and maximum population, and they can provide Helper
classes. Also, interfaces can be part of a hierarchy, like classes.

11.5   The `RealObjectsServer`

The main purpose of the `RealObjectsServer`, or ROServer, is the collection
of RO clients and the implementation of methods to *serve* their invocations.
By design, any ROApp should extend the ROServer to manage a number of remote ROs. Note that the server is coded against `RemoteObjects`,
not `RealObjects` nor their extending implementations. The extending real-
object application should be able to override the methods to modify or
enrich functionality according to its "rules." Extending classes can be seen
as programming in layers, and it makes sense to create a class for each
semantic layer early in the design and modeling process.

Note that the ROServer does *not* extend the `UnicastRemoteObject` although *it is* a RMI server. RMI allows any Java object tagged with
a `Remote` interface to be exported, in this case the ROServer. Subtyp-
ing `UnicastRemoteObject` has the advantage over `exportObject()` in that
`hashCode()`, `equals()` and `toString()` are pre-implemented. These methods are more useful for the `RemoteClient`.

The server engine's `main` method reads (or creates) the property files and
sets the properties in the software. With the properties `host port name`,
the server can be exported with the methods of the `roaf.util.RMI` class:

```
exportROServer( String host, int port, String name )
```

Note that the server is already extended to a concrete application:

```
public class LondonChase        // RealObjectsApplication
     extends RealObjectsServer
  implements RemoteObjectsApplication // RemoteLondonChase
```

`LondonChase` *is a* real-object application and for clear semantics the interface is named `RemoteObjectsApplication`. Once the server is instantiated,
exported, and published, the RO clients can look it up and apply methods
of this (composed) remote interface. According to the vision the clients
run independently of the server process, and for timing reasons, it makes
sense to split the log on into two phases: `identify` and `enter`.

11.5.1   Client-Server Dialog

Client call: `ServerInfo identify( RemoteObject client )`. In phase one
the client has to identify (register, authenticate) itself. Think of a `Real
Airplane` approaching the RGB scenario (page 77) on its way to the Mu-
nich airport. With the plane traveling at some hundred kilometers per
hour, it will traverse the RGB bounding box in a few minutes. So, if the
plane shows up in the scenario, it should identify itself clearly before cross-
ing the scenario's boundaries. In this case, the identification process could
sample the airplane's direction and speed in order to be ready to handle it
as soon as it flies over the moving motorcycles and calculate its shadows
from the known position of the sun.

By invoking the `identify` method, the client transfers its remote reference `RemoteObject` to introduce itself to the server. The server can store and
use this remote control to observe the client's motion and determine the
speed and direction. Since the ROServer is developed with *one* concrete
application, it can not know all of the scenarios. Nevertheless, it should
fit all scenarios in the long run. Based on the remote object's methods,
the server can check synchronicity. Another test could compare client and
server metrics with the `distance` method.

If you have already set up the application (see Section 13.2.1), you can look at the player identification sequence. The `RandomPlayer` is an abstract `LCPlayer`, which is a `RealObject`. The identification can be found in `LCPlayer.identifyPlayer()`. The server implementation is in `RealObjects Server.identify( client)`.

The (to be returned) `ServerInfo` is a serializable exchange object between client and server. It is a record of well-defined server responses to interactive actions to let the client know what has happened, if the action was successful, etc. In addition to the predefined `int` values, the server can add a human-readable message `String`. `ServerInfo` can be a useful and flexible return value for most client invocations on the server.

Server callback: `Object getIdentity()`. Many server applications use a server-side generator to provide a client ID. In the case of the ROAF, each RO is required to provide its own unique global identity. In the real world, people are identified by passports; computers can be identified by a MAC or IP address; more and more consumer articles (and animals) are identified by transponders like RFID tags, etc.

To leave the choice open, the `java.lang.Object` will be used as the most general (Java) identifier. This general ID can be cast, if client and server agree on a more concrete identifier. In the long term, a digital fingerprint like a X.509 certificate should do the trick (and can be approved with ROApp signatures).

The inner class `RealObject.RemoteClient` implements `getIdentity()` required by the `RemoteObject` interface *without* declaring a `RemoteException`. The reason for this is that any remote object is *exported* before it can be addressed. Internally, the RMI technology creates a hidden object with all remote methods throwing `RemoteException`s. Generally, if an exception is thrown in a base class method, the subclass method can throw the same exception, a more specific exception type, or no throw's clause at all. Nevertheless, the constructor needs to throw the remote exception, and it's good practice to add the throw's clause.

With the `client` reference, the server *could* immediately invoke the clients methods:

```
Object ID = null;
try  { ID = client.getIdentity(); }
catch (RemoteException e) { return null; }
return ID;
```

A closer look with a simple sequence diagram reveals a typical server challenge:

```
                           exchange
        server machine      Objects        client machine
        --------------      --------        --------------
                           <- client -     server.identify( me )
    client.getIdentity()  ----------->     wait for keyboard input!
      approve / deny ?     <- Object -      return
          return           - feedback ->   ENTRYAPPROVED ?
```

After one client has passed its remote reference, the server calls back and
tries to retrieve an ID via `client.getIdentity()`. Technically, the client
invokes a method on the remote server, which invokes a method on the
remote client. Although the nested construction works, it can be tough to
debug. The actual problem arises, if the client response is too slow.

The sequence diagram indicates that the identification could be provided
via keyboard on the client side. In case the client operator is distracted
by something the server keeps waiting for the ID to be returned,[1] and
the server method cannot continue or can get confused with other clients.
In general, it is unwise to call a method from a thread, if the method is
open for implementations like a method from an interface or an overriding
method. The programmer simply can not make any assumptions on the
client method's return.

Server programming is about handling many client threads at a time,
while managing resources and balancing the load. The client call and the
server callback have introduced this issue; general ways to handle the challenge will be discussed in Section 11.7.

To avoid an identification issue, a simple `Map` is created to make use of
Java standard implementations. Java supplies a runtime (!) `hashCode()`
and an **equals()** method for the **RemoteClient** and a `HashMap` to detect
duplicate clients:

```
private   Map<RemoteObject,??> connectedClients
= new HashMap<RemoteObject,??>();
```

The `??` indicate the need to identify server-side objects relating to each
client object. The map `connectedClients` collects unique `RemoteObjects` as
mapping keys. The `RemoteObjects` are local representatives of the client
inside the server environment—`ServerObjects` or SOs. A SO is the counterpart or reflection of any `RemoteObject`, and the server is the mirror!

11.6 SOs: `ServerObjects`

Each SO is a dedicated server process for every RO. The `ServerObject`
provides an individual programming environment on the server side for each
remote client. For the ROServer, it is easier to deal with local objects and

---

[1]Although **void** methods don't return a value, even they do return!

delegate individual responsibilities to sample information and propagate and store warnings; if a connection is lost the SO can try to call back the related RO, etc.

Before `identify( client )` returns a `ServerInfo()` response the server creates the "client record." Instead of a normal constructor,

```
new ServerObject( client, ... )
```

the method

```
ServerObject createServerObject( RemoteObject client )
```

has the sole purpose of creating a server object. If the construction of a server object were hard-coded inside the server, the application would be restricted to the pre-implemented server object; the protected method can be overridden to create any object extending the server object! This will become clearer when the server is extended by an application (see page 173).

Server profile. The two phases `identify` and `enter` of the login process are more than the usual ID and password combination on most websites. This is due to the fact that a profile stored on a website is persistent and static, while a RO is alive. Each ROServer should be able to keep *dynamic profiles* of its RO clients.

From the developer's perspective, inheritance is another term for reusability in similar environments. One intention of the `ServerObject` is to store the profile, history, and behavior of the RO client. By inheriting

```
ServerObject extends RealObject
```

the SO automatically gains the ability to record, store, and play back a live trace. The live trace can be time-stamped with the server time and can be played back in server space in a synchronized manner with the other RO client records. Also, the built-in ability to be published as a `new` `RealObject`, reachable as `RemoteObject` from other participating objects and servers, is part of the server design process. This also makes a lot of sense from the software maintenance aspect

The method

```
connectedClients.put(client, candidate);
```

maps the unique client RO to its dedicated SO.

The next step is the actual identification to become a candidate of the server environment to `enter` the scenario.

11.7   Server Tuning

A server machine usually has massive CPU power to handle a large number of synchronous client accesses. Nevertheless, the server has to make sure that the server-side objects accessed and modified by client calls are served in an ordered manner to ensure that the `ServerObjects` are always in a clean state. Since client method invocations can take their time to return, the server should synchronize its services. Some Java applications add (un)locking methods for the client. This can be risky and will not be used for RO clients.

11.7.1   Threading (in a Nutshell)

For thread implementations, the following aspects need to be considered:

1. Identify the objects that can be accessed by concurrent threads. Keep in mind that each remote client represents an independent thread.

2. Ensure that a thread can not modify an object while another thread is modifying it. Registering a client should be *finished*, before another client can *start* registering. This can be achieved by using a `synchronized( object ){ ... }` block. The JVM ensures that only one thread can access the synchronized object. When the code is entering this block, other accesses to this object are blocked/locked and unblocked/unlocked upon leaving. A lock *only* controls access to the synchronized code.

3. Introduce `synchronized` methods in the object being accessed concurrently which needs internal book keeping. A `synchronized` method automatically uses `synchronized( this )` to block `this` object with the method. The object should introduce object *states*, which can be used in `synchronized` blocks to control the program flow. Other threads can use the object states to control their flow. Method names like `.waitToStart` and `.waitToEnd` should indicate that these methods implicitly call `Object.wait()`. Method pairs can imply do and undo operations for object states. These methods support the coordination of many threads and leverage (make use of) the internal `java.lang.Object`s states and triggers controlled by an internal scheduler.

Since the construction of a thread is a complex JVM operation every server should have a concept to deal with many threads.

11.7.2   Handling Multiple Server Threads

If the ROServer's `identify` method were `synchronized`, one client could
block the identification of other clients. One solution to this timing problem
is to create a deferred thread to take care of the remote client call and
immediately continue inside the `identify` method.

The `DeferredThread` *could* be added to the server (object) as a private
class:

```
private class DeferredThread extends Thread
{
   DeferredThread( Runnable r ) {
      super(r);
      setDaemon(true);
   }
   public void run() {
      try { Thread.sleep(1000); }    // wait a second ...
      catch (InterruptedException e) {}
      super.run();            // ... and execute client call
   }
}
```

Setting the thread as a daemon is a precaution to allow the server to shut
down while clients are still trying to identify themselves.

By using the thread to call the client method, the server's identification
method gets rid of the client call and can return its response. After a
second, the client can be identified for any duration:

```
DeferredThread getClientID = new DeferredThread(
   new Runnable() {
      public void run()    // wrap client call in thread
         { candidate.getIdentity(); }});
getClientID.start();      // submit client call ...
   ...                    // ... and immediately continue
```

The performance drawback of this approach is the creation of a new
thread for every method call. Instead of creating a deferred thread the Java
library has a `java.util.Timer`, a facility for threads to schedule tasks for future execution in a background thread as well as a `java.util.TimerTask`, a
task that can be scheduled for one-time or repeated execution by a Timer.
The advantage of the timer utility is the one-time creation of a thread,
which will stay alive as long as the timer is working. This built-in mechanism could be used as the base element of a more complex *job scheduler*
to take care of many server processes. For a small number of clients and
methods, this solution works fine. A server developer has to be aware
that there are probably more threads running than he can actively know.
Obviously, a GUI, like Swing, uses internal threads; RMI uses threads to
marshal (pack) and unmarshal (unpack) objects, etc.

Since the ROAF aims at systems of larger scale, the server should use a more general approach to set off a large number of client calls with a decent response time. A *worker queue* or *worker thread* is a general term for a sequential chain of tasks usually worked off in one thread. This could be useful for the sequential order of rounds and moves in a board game. Another approach is a *thread pool*, which creates a (given) number of (worker) threads waiting to process a number of client calls in parallel. This solution is faster than a single worker queue, although the calls are processed without a specified order.

Java 1.5 has integrated the `java.util.concurrent` package to decouple task definition and task execution for general server programming.

THE JAVA TUTORIAL > ESSENTIAL CLASSES > EXECUTORS

> ...there's a close connection between the task being done by a new thread, as defined by its Runnable object, and the thread itself, as defined by a Thread object. This works well for small applications, but in large-scale applications, it makes sense to separate thread management and creation from the rest of the application. Objects that encapsulate these functions are known as executors.

The `java.util.concurrent` package basically separates the creation and execution of tasks with the `Executor` and `ExecutorService` interfaces. A *task* is a well-defined unit of work like a server request, ideally independent, while interfaces separate *what* and *how* of the general *consumer-producer* pattern.

By programming the server application against these interfaces, the actual scheduling and load balancing can be chosen at deployment time. The choice depends on hardware considerations, like the number of CPUs, and expected throughput and responsiveness of the application. Today even most personal computers operate on multiple processors. With the concurrent package, the service can be added to the ROServer as a member:

```
private int nrOfParallelThreads = 5;
protected ExecutorService executorService
    = Executors.newFixedThreadPool( nrOfParallelThreads );
```

The fixed thread pool is supplied as a default service for testing on the programming environment while the extending ROApp can override the initial choice with an adequate implementation. The choices to scale the server at deployment time can be a sequential (un)bounded worker queue or a parallel thread pool of queues. For sequential and unbounded task processing the choice could be:

```
protected Executor serverQueue
      = Executors.newSingleThreadExecutor();
```

The thread pool "creates an Executor that uses a single worker thread operating off an unbounded queue" [see javadoc, `Executors > newSingleThread Executor`] or a thread pool with a fixed number of threads:

```
protected Executor serverQueue
      = Executors.newFixedThreadPool( nrOfParallelThreads );
```

The identification of the player client could be added with

```
Runnable getClientID = new Runnable() {
      public void run() { candidate.getIdentity(); }};
serverQueue.execute( getClientID );
```

This construction is a bit simpler than `DeferredThread` method used earlier. A deferral could be placed inside the `run()` method,

```
...
  public void run() {
      try { Thread.sleep( 1000 ); } // defer call
      catch (InterruptedException e) {}
      candidate.getIdentity();
  }
```

to slow down a single worker thread, while a thread pool can still accept other requests. Probably the best way to do this is to use a scheduled thread pool

```
protected ExecutorService serverQueue =
    Executors.newScheduledThreadPool( nrOfParallelThreads );
```

A scheduled thread pool "creates a thread pool that can schedule commands to run after a given delay, or to execute periodically" [see javadoc, `Executors > newScheduledThreadPool`].

For the identification, the `Runnable` interface was used, which is usually associated with `Thread`s. Actually, it is not tied to threads, so it's fine to use it for executors. The `Runnable.run()` method imposes limits on the implementation as it can not return a value and can not throw an exception.

These limits can be overcome by using the concept of the `Executor Service`, `Callable`, and `Future` classes. Just like `Runnable`, the `Callable` interface describes tasks to be `executed` or `submitted`. The callable interface makes use of generics (which weren't available when `Runnable` was introduced) to supply return value and exception with the implementation:

```
public interface Runnable    { void run(); }
public interface Callable<V>
                { V call() throws Exception; }
```

The main idea of a thread pool is to un-synchronize a large number of tasks with different durations. Consequently, the executor framework decouples a *method call* from a *method return*. From a logical point of view, an application submits a task and can check on the result some time in the future. The construct to fetch the result is the `Future` class.

The client call can be coded as

```
Callable<Object> getIdentity = new Callable<Object>()
  { @Override
    public Object call() throws RemoteException
    { return remoteObject.getIdentity(); }};
```

and submitted with

```
Future<Object> taskResult
          = executorService.submit( getIdentity );
```

Then, the executor services' `submit` method for `Callable` tasks organizes a thread. Since the result is returned asynchronously, the method returns a `Future` immediately. The application can use the waiting time to do something else. The `Future` object supports different strategies to block until the result is available, to wait a given time span, or to check on the result or to cancel the task.

`java.util.concurrent.Future<V>`

A Future represents the result of an asynchronous computation. Methods are provided to check if the computation is complete, to wait for its completion, and to retrieve the result of the computation.

- **`V get(long timeout, TimeUnit unit)`**
  Waits if necessary for at most the given time for the computation to complete, and then retrieves its result, if available.

- **`V get()`**
  Waits if necessary for the computation to complete, and then retrieves its result.

- **`boolean isDone()`** Returns true if this task completed.

- **`boolean cancel(boolean mayInterruptIfRunning)`**
  Attempts to cancel execution of this task.

- **`boolean isCancelled()`**
  Returns true if this task was canceled before it completed normally.

The `ExecutorService` adds life-cycle support to shut down the executor framework and is compatible with the `Executor` interface. Developers

should experiment with different implementations as the execution policies
offer a wide range of transaction management configurations.

11.7.3   The Remote Concept for Server Objects

The `java.concurrent` package provides concurrency on a higher (i.e., server)
level than the low-level methods in `Objects` and `Threads`. For the application
designer, it would be even more pleasant to have *one* concept for all remote
client calls pre-implemented in the `ServerObject`. For the actual scenario,
all SOs are treated as local `RealObjects`—which, in fact, they are.

The executor service is instantiated in the ROServer and can be referenced by all SOs. A different approach to provide the service could be a
`static` executor service inside the `ServerObject` class.

First, `Callable` is restricted to remote calls with

```
protected interface RemoteMethod<V> extends Callable<V>
    { V call() throws RemoteException; }
// read: "RemoteMethod.call throws RemoteException"
```

The interface is placed in `ServerObject` to be used by all extenders. Generic
placeholders indicate that the actual values need to be supplied with the
implementation and before compiling. Therefore any method using the
`RemoteMethod<V>` has to be coded as a class, which can be implemented
and instantiated for a dedicated remote method. The class `RemoteCall<V>`
wraps all executor options to hide their complexity from the external client.
The concept does not need to distinguish methods with or without return
values as the `<V>` includes the class `Void` as a virtual return value with a
precise time of return.

The constructor is used to store the `RemoteMethod` for the `RemoteCall`
submission:

```
protected class RemoteCall<V> {
   private RemoteMethod<V> remoteMethod;
   public  RemoteCall( RemoteMethod<V> rm )
                    { remoteMethod = rm; }
   ...
```

Then, the different `Future` strategies are handled in a single method:

```
public V get( int waitMillis )
{
   Future<V> remoteTask
             = executorService.submit( remoteMethod );
   try {
      if ( waitMillis == -1 )      // wait for return
         return remoteTask.get(); // can block!
      else
         return remoteTask
            .get( waitMillis, TimeUnit.MILLISECONDS );
```

```
        }
        ...
```

By explicitly defining the maximum waiting time (in milliseconds), the
method call can *always* return within the given time window. Otherwise,
it will throw a `TimeoutException`. If the method does not return, or throws
a `RemoteException`, additional handling is necessary:

```
        ...
        catch ( InterruptedException ie) { /* TODO */ }
        catch ( ExecutionException   ee) {
          if ( ee.getCause() instanceof RemoteException )
                /* TODO */
        }
        catch (CancellationException ce) { /* TODO */ }
        catch (Exception e)
                { /* TODO: unexpected Exception ... */ }
//      finally { /* TODO */ }
        return null;
    }
```

Although `RemoteMethod.call` can only throw `RemoteExceptions`, the exception has to be picked up via `ExecutionException.getCause()`, which is part
of the executor concept. If a task throws an exception, `Future.get` wraps
it in the `ExecutionException`.

In case the external client chooses to ultimately wait for a method return,
it should get the chance to cancel the task explicitly, i.e., to shut down the
application or server:

```
    public boolean cancel( boolean mayInterruptIfRunning )
    {
        if ( remoteTask == null ) return false;
        return remoteTask.cancel( mayInterruptIfRunning );
    }
```

`RemoteCall<V>` supports waiting for client responses for a maximum time.
If there is no return (value) after the given time span, the calling (SO-)
method can implement a fall-back strategy.

Using `RemoteCall.submit()`, the `RemoteMethod` can be submitted exclusively. In case the client method is void and/or the server does not bother
to wait for its return, there is no further need to use `get`. Nevertheless, the
result (or exception) can be retrieved at a later time via `get`.

Using the remote concept. After this excursion on threading concepts,
the developer needs a recipe to apply the chosen concept for his remote
method calls. In terms of program flow, the text has not reached the end
of `ROServer.identify` yet! The client has been mapped to the `connected
Clients` and is then asked to identify itself with a single line of code:

```
    Object ID = candidate.getIdentity();
```

To get the remote clients identity, the concept is applied inside the SO and overrides the platonic RO method. First, the remote method to call the remote client is implemented and externally hidden as private:

```
private RemoteMethod<Object> getIdentity
  = new RemoteMethod<Object>()
    { @Override
      public Object call() throws RemoteException
          { return remoteObject.getIdentity(); }
    };
```

Then, the actual method can be published with

```
final public Object getIdentity()
{
      RemoteCall<Object>
      remoteCall = new RemoteCall<Object>( getID );
      Object remoteID = null;
      try  { remoteID =
              remoteCall.get( maxWaitForRemoteObject );
//            validate remote value ...
      }
      catch (TimeoutException e) {
//         create fallback value ...
//         remoteID = fallback;
      }
//    log ...
      return remoteID;
}
```

The `RemoteMethod` implementation can be created inside the method on the fly (inline) or, separately, as above. The latter has slower performance, but can be convenient if the remote method needs to be called with parameters. Both strategies have access to the environment (i.e., the `ROServer` class).

The `/* TODOs */` in the `RemoteCall` listing can be filled in step by step, when testing the code. The primary concern is the handling of `Remote Exception`, which indicates that something is wrong with the client-server connection. The server can be informed asynchronously, while the scenario should continue with a plausible fallback value. From there on, the server logic can decide what should happen, for example, stop the game in progress, if a minimum number of players is required.

Client server propagation: `report2client`, `report2server`. When the server logic requires a RO position, the SO has to retrieve this information remotely and propagate it to the application process; vice versa the SO informs the RO of server events.

The method invoked on the server should be void to indicate a dead end. Although no value is returned, the programmer can place a number of internal methods for asynchronous actions on the running application.

As any object, the application can work with states reflecting the current status and these can be influenced, as, for example, setting the status to a radical GAMEOVER state.

11.7.4   A Fallback Scenario for getGPSinfo().

So far, the RealObject was only used in playback mode. If the RealObject is actually running on (or receiving from) a (Java-enabled) GPS device, like a modern smart phone, the virtual GPSunit is actually fed from this device via hardware interfaces (drivers). On the server side, the ServerObject can not distinguish the client's mode and simply receives the information as if it is all coming from a GPS device.

The SO remote concept provides a notch for a "fallback strategy"[2] in case the connection gets lost similar to a real GPS device. The effects can be seen on GPS traces with bad positioning of way points due to obscured satellites. A validation against a street geometry always needs a certain tolerance. An application could also implement a tolerance to connected RO clients, which again is a realistic scenario. If you travel with your smart phone, you can record a trace offline and connect to a server every once in a while to submit (fragments of) the trace or the current position—while remaining an actor in a scenario.

This constellation can be implemented easily. The SO applies the remote method getGPSinfo to check where the client is and setGPSinfo to synchronize SO and RO location. If the RO is offline, the SO can guess the RO position by extrapolating the current position from the last-known position, direction, and speed and return the SO.GPSinfo to the server logic. This fallback strategy is plausible and nothing more. It is similar to letting go of your steering wheel—in a curve! The car is driven by inertia and hits anything in the way.

Remote concept catch. All in all, the remote concept is very convenient. Nevertheless, it does not cover every concurrent case, and the programmer has to be aware of typical pitfalls. No matter what kind and what size of executor service you might apply to your application, keep in mind that threads are always competing due to a restricted number of CPUs.

The sequence diagram on page 149 indicates that the server.identify method can never end without a prior return of client.getIdentity. During development, one should always be aware of sequential methods that rely on one other. Since the executor framework has different implementations, a simple working queue can get problematic, if a method is submitted

---

[2]To satisfy the RO vision (page 18): every RealObject must provide a location . . . at any time.

and a sub-method is scheduled right behind it. Then, the method is waiting for the sub-method, while the working queue is waiting for the method to return.

With the remote concept, a dead lock will at least be timed out and provide hints where to look for the nesting. In the end, a concurrent logic may have to be introduced by using `synchronized` methods to block a method until it is its turn. The higher level concurrency package can never replace logical concurrency of collaborating objects.

With a lot of thread theory, the architect has established a general approach to handle client threads hidden to the application developer. The two sample implementations `getIdentity` and `getGPSinfo` in the SO will automatically be tested in the development process and provide a shell to add code according to experience gained. The reader is advised to go through the application implementation in the next chapter to get a better grip on the concept with an external view on the SO implementations.

11.8   The `RealObjectsBox`

The identification process of the client with the server creates a `Server Object` for each `RemoteObject` to make it eligible to enter the scenario. On the one hand, the SO serves as a record of the RO's behavior and, therefore, needs to be stored and referenced in the server environment. On the other hand, the server scenario is composed only of SOs.

Before even going into details, it is obvious that the SOs are vital for each scenario. This observation is sufficient for an OO architect to create a new class `RealObjectsBox` or `ROBox` to handle the details to come. Architects usually don't consider furniture when they construct a building—only functionality. With a functional object, the architect is able to fill in details while controlling access.

What should the `RealObjectsBox` do? Briefly, the `ROBox` controls the RO population of a scenario.

After a client is identified, it should become a `candidate SO` authorized to enter the server scenario. The SO is only constructed successfully, if the remote client can be identified. The `identity` object should be unique for a worldwide application. Nevertheless, it is (yet) only an `Object` and can not identify the client instance. According to the ROAF Turing test (page 18), the server has to observe clients for a while to validate plausibility.

After a thorough identification (in the overriding ROApp), the SO is related to its `identity` object:

```
synchronized boolean
          putCandidate( ServerObject newCandidate )
{
  Object ID = newCandidate.getIdentity();
```

```
    ...
    candidates.put( ID, newCandidate );
    ...
}
private Map<Object,ServerObject> candidates
  = new HashMap<Object,ServerObject>();
```

The candidate collection maps the intrinsic ID to its SO and reduces to server logic to the handling of ID keys to interact with a client. The synchronization of the ROBox is vital to avoid two clones running into a conflict; the rule is first come first serve. Although the utility class Collections provides methods to make collections synchronized, the architect decides to synchronize the access *on the* ROBox wrapping candidate and actor's collections. (Actors are the ROs that have entered the scenario to become part of it.)

After creating a server object, mapping it to the identity object, and putting it into the candidate list, the identification process is completed and the candidate can get ready to enter the scenario and become an actor (in a few pages):

```
synchronized boolean putActor( Object ID ) { ... }

private Map<Object,ServerObject> actors
  = new HashMap<Object,ServerObject>();
```

Note that putActor is using the ID while putCandidate is using the SO as parameter. These semantics imply that an actor has to be an identified candidate to be able to enter the scenario. Depending on the server application, the remote client might have to be recognized by another identification process.

With a simple integer identifier, the client programmer could apply a hack: use one client to identify and enter with another client with the same ID. Of course, the server should be able to expose technical hacks.

A "nice-to-have" is a dynamical spatial index for all actors. That way any client connected to a scenario could restrict the number of actors by the bounding box relevant to himself.

11.9 ROAdio Broadcasting

At this point the client server communication via RMI should be clear. Server and client have a remote reference to each other and can retrieve information as needed. The executor framework decouples the server response to various clients.

The Bingo application (see page 144) suggests another type of communication:

> THE JAVA TUTORIAL > PUTTING IT ALL TOGETHER
> > COMMUNICATION BETWEEN THE GAME AND THE PLAYER
>
> ... the Game and the Players have two different forms of
> communication and use two different mechanisms for them.
> First, the Game broadcasts status information to all the Play-
> ers over UDP. Second, the Player makes individual requests of
> the Game via RMI.

RMI communication can be compared to a phone call between two directly connected individual devices, while broadcasting via a multicast
socket is similar to a radio station transmitting information to anyone lis-
tening. In the context of a ROApp, the server could choose *not* to communicate through the RMI connection, but simply inform a "public audience"
about the ongoings. Broadcasting relieves the server, since it doesn't have
to wait for a method to return. And it can also relieve the client as it can
choose to use the information, simply ignore it, or not even listen to it in
the first place.

Again, the architect has to make a decision to add another communication technique or keep the focus on the overall progress. In order to make a
qualified decision, the analysis process should be pointed out regardless of
the choice. This section represents a protocol of the analysis phase, thorough enough to pick it up in a later stage or cycle. For a global real-object
framework composed of ROServers and ROApps, it makes a lot of sense to
broadcast general information and give the clients a chance to see what's
going on before they choose to participate in a scenario. Servers may also
want to listen to the ROServer layer and choose to collaborate with it, i.e.,
use its server objects as a trusted source.

How It Works

Broadcasting is accomplished using User Datagram Protocol (UDP) and
does not require point-to-point connections. Rather, it sends packages,
packets, or datagrams and does not have control over their arrival. The
methods `java.net.DatagramPacket` and `DatagramSocket` implement UDP.
The `DatagramPacket` can be simplified to a byte array containing any information the sender (packing the packet) chooses to send. The receiver
of the packet has to know how to unpack the information and what to do
with it. Like (re)constructing a java `Object`.

The class `bingo.shared.PlayerRecord` has three steps:

```
public class PlayerRecord {
// content of the class:
```

```
    public int ID = -1, numCards = 0, wolfCries = 0;
    public String name;

// 1. regular construction in the java environment:
    public PlayerRecord(int ID, String name, int numCards) {
        this.ID = ID; this.name = name;
        this.numCards = numCards; this.wolfCries = 0;
    }

// 2. sender disassembles object to a byte array:
    public byte[] getBytes()
    {
        byte[] numbers
            = { (byte)ID, (byte)numCards, (byte)wolfCries };
        byte[] nameBytes = name.getBytes();
        byte[] answer
            = new byte[numbers.length + nameBytes.length];
        System.arraycopy
            (numbers, 0, answer, 0, numbers.length);
        System.arraycopy
            (nameBytes, 0, answer, numbers.length,
             nameBytes.length);
        return answer;
    }

// 3. receiver reconstructs new object from a byte array:
    public PlayerRecord(byte[] b) {
//      single byte fields
        this.ID = (int)b[0]; this.numCards = (int)b[1];
        this.wolfCries = (int)b[2];
//      bytes to a String field
        byte[] nameBytes = new byte[b.length-3];
        System.arraycopy
            (b, 3, nameBytes, 0, nameBytes.length);
        this.name = new String(nameBytes);
    }
}
```

The listing points out the fine print of UDP communication: server and
client have to agree precisely *how* to disassemble and reassemble the transmitted information and to *what*. In other words, they have to come up
with their own protocol. With the built-in protocol of RMI and serializing
objects, the programmer only has to know *what* to send and receive. Note
that the PlayerRecord is not tagged to be Serializable.

The byte array is packed into a DatagramPacket for the actual broadcasting:

```
class SocketGate
{
    ...
    void sendPlayerStatusMessage(PlayerRecord p)
    { sendBytes(p.getBytes(), playerListeningGroup); }
```

```
      private void sendBytes(byte[] data, InetAddress group) {
        DatagramPacket packet
        = new DatagramPacket(data, data.length, group, port);
        try { socket.send(packet); }
        catch (java.io.IOException e) { ... }
      }
  }
```

In radio or television terminology, the `SocketGate` represents the broadcasting station, the `MulticastSocket` the broadcasting technology, and the `InetAddress`(es) the channel(s), which are all set in the constructor:

```
SocketGate () throws java.io.IOException
{
  sender  = new MulticastSocket  ( port );
  station = InetAddress.getByName( channel );
}
```

The server logic can broadcast an existing player record with

```
socketGate.sendPlayerStatusMessage( playerRecord )
```

The client, or rather receiver, has to implement one socket for each channel

```
socket = new MulticastSocket( port ); // turn tuner on
socket.joinGroup( channel );          // dial a channel
// vice versa end with:
// socket.leaveGroup( channel );
// socket.close();                     // turn tuner off
```

to be able to receive datagrams with

```
byte[] buf = new byte[256];
DatagramPacket packet = new DatagramPacket(buf, 256);
try {
  socket.receive(packet);
          // method blocks until a datagram is received
  byte[] rcvd = packet.getData();   // create byte array
  final PlayerRecord p
  = new PlayerRecord(rcvd);// and player from byte array
  anyMethod( player ); // propagate to local application
} catch (IOException e) { ... }
```

Note that first the datagrams *data* has to be retrieved, before the data is actually turned back into useful *information*.

For completeness, it should be mentioned that the Bingo application places `socket.receive` in a loop for repeated receptions, wraps the propagation in a thread, and submits it to the Abstract Window Toolkit (AWT) event-dispatching thread. The Bingo tutorial also describes the wins (shared components update via broadcasting) and the optimistic handling of primitive values. And, maybe the most interesting aspect: a broadcast datagram (and time) can be used to trigger RMI communication. For more details on networking see the Java Tutorial.

THE JAVA TUTORIAL > TRAIL: CUSTOM NETWORKING

> The Java platform is highly regarded in part because of its suitability for writing programs that use and interact with the resources on the Internet.

In the end, the architect decides that it is useful to give all operators of an application's remote objects the opportunity to chat. The server can receive a `String` and propagate it via a common channel. This keeps the protocol really simple and reliable, since the standard `String` implementation provides methods to convert to bytes and back. This way the server is the third party to receive (censor) and broadcast the information to all chat clients. Collaborating remote objects can agree on their own secret protocol to exchange information—encrypted for unwanted listeners.

Chapter 12

Rolling out a ROApp

12.1   Introduction

After a rough construction of the ROServer, the architect directly adds the ROApp layer as indicated by the four-layer architecture in the previous chapter (page 145). Server and application layers represent the technical and semantic aspects, respectively.

The mission of this book is the creation of an executable *demonstration prototype*, which is equivalent to establishing a first cycle. This chapter ends with the barebones of a game. The reader can experience his own life cycle development by playing, adding rules, implementing details, and playing against other readers and/or computers.

12.2   LC: The Game Scenario

The architectural approach is different than the developers', who has to add the details later. The architect creates the four components representing the layers as early as possible and *then* adds functionality (with preliminary prototype implementations) step by step. Every concrete application can introduce new technical challenges, which should be shifted onto the server on a long-term basis. From the object-oriented view, the application extending the server represents *one* server object for a specific application.

Developers like games and board games, in particular, have clearly defined, discrete rule- and game sets. After implementing these, the programming challenge is the automation of players with programmed intelligence for decision making.

We will use the board game *Scotland Yard* for our example.[1] This game was chosen, since the game board is basically a city map of London reduced to a network with 199 stations and three map features `cab bus sub` similar to the Germany map already used for the `Navigator` on page 129. Since the

---

[1]The current publisher is Ravensburger; a previous US edition was published by Milton Bradley. See `http://en.wikipedia.org/wiki/Scotland_Yard_(board_game)`.

game rules should be reduced to a minimum, in keeping with the ROAF paradigm, the revised game is called `LondonChase` or LC.

Although game rules and game sets represent programmatic constraints, they should be kept flexible for the number of players, etc. A good way to achieve this is by distributing them over the properties files for server, application, and client software as predesigned in the last chapter. Obligatory default values can be hard-coded in interfaces. The architect abstracts the actual game by creating a skeleton with enough code to begin a game and run it for a few turns.

After reading this chapter, the reader should be able to enrich the game, or even better, create his own version of the game. With the know-how about processing digital maps, the game can easily be transferred to any other town.

12.2.1 Language Analysis of the Game Rules

A player, alias Mister X, is moving along a network of taxi, bus lanes, and subway lines. The game master reveals his position (current station) every x rounds. Additional players, called Detectives, are moving along the same net to chase Mister X and catch him at a station. The detectives are able to communicate with each other in order to develop a strategy to track down Mister X. The game is "digitized" by game rounds to make initial development easier. Every player makes one move per round in sequential order. Mister X wins, if he is not caught before the final round is completed. These simple rules are sufficient to implement a basic game application.

Once the initial game is implemented, the reader can add items from the original game, which uses a pool of taxi, bus, and subway tickets for each player depending on the total number of players. The original game board also has a boat feature for ferries on the Thames.

With every move, a player has to use one ticket according to the network link feature used. The player can only make a particular move, if he has a ticket for it. Naturally, this affects the strategy. Mister X has additional options like double move and black (any feature) tickets. The detectives get to see the tickets used in every round. Similar games, like New York Chase have additional features like helicopters, or they allow players to drop barriers[2] forcing other players to route around them.

12.2.2 Game Place and Time

Every ROApp should be restricted to a bounding box representing a part of the real world. In the case of `LondonChase`, the box is defined as a part of London. Initially, the game time does not have to be set to real time. In

---

[2]Real-world barriers: `wiki.openstreetmap.org/wiki/Map_Features#Barrier`

later development cycles, the real public-transportation schedules could be added to make the game more realistic, and *then*, the scenario can become a candidate for a real-world application. Actors (and their developers) could access schedules via the internet as a third-party source. Players in the United States would have to play with the adjusted time from London.

As a general design rule, a ROApp should *not* be set to real time to indicate that it is *not* trying to reflect the real world. Therefore, the default ROServer constructor is standardized to the beginning of this millennium, which is convenient to tell the server uptime. With `name host port` it is easy for RO clients to look up a server and find out the where and when of a scenario:

```
String rmiURL = "rmi://" + host + ":" + port + "/" + name;
Remote connection = RMI.lookupRMIURL( rmiURL );
RemoteObjectsServer
          roServer = (RemoteObjectsServer) connection;
RemoteObjectsApplication
        roApp = (RemoteObjectsApplication) connection;
serverBox  = roApp.getServerBox();
serverTime = roApp.getGPSinfo();
```

12.3   Server Architecture

Although the ROApps support the ROAF, there is no `RealObject Application` class in the `roaf.roa` package! There are simply too many (actually an infinite number of) ROApp candidates in the real world. *You* decide which part of the world you would like to model, and how. At development time, the ROServer can not account for all future ROApps. Yet the ROServer is the place to implement controlling mechanisms (supported by scientific laws) as they occur.

12.3.1   Components and Responsibilities

In the previous chapter, a rough client-server architecture was developed. With the concrete application LC, the architect can refine the right-hand side of the four-layer architecture shown in Figure 12.1. The language analysis of LC (see Section 12.2.1) provides the basic components for the scenario. The components of Figure 12.1 provide an overview of server and application responsibilities as well as subcomponents accounting for separate tasks. On the left-hand side, we see that the server is open for remote client objects like cars (taxis), busses, trains (subways), and houses (or stations). The client *candidates* are collected in the `ROBox` after they have identified themselves. Next the server can `class`-ify the clients to familiar classes. While the ROServer code should only use the (abstract) base classes defined in the `roaf.ros` package, the ROApp can use the final

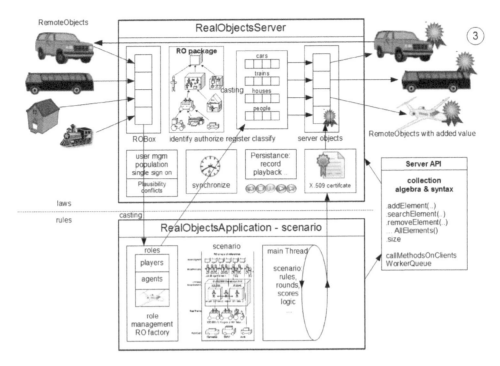

Figure 12.1. The server consists of a ROServer (library layer) and a ROApp layer. Each layer has different components with individual responsibilities.

implementations and (try to) upcast accordingly. In the long run, many ROServers will represent an entry into the ROAF and a single sign-on mechanism can spare the clients from repeated identification processes. One could say that one server can recommend a client to another server.

A client is authorized by the application to enter a scenario. In the game *London Chase*, the client has to be a valid `LCPlayer` (to be introduced on page 175) in a predefined role. The application can restrict the scenario to a time frame by waiting for the minimum number of players before starting a game. The LC game will implement a `GameThread` to run a given number of rounds and sequentially collect the players moves. There should be a recording mechanism in the game so it can be replayed at the end to check that no one has cheated.

The ROServer and ROApp should split up their responsibilities as follows. The ROServer

- handles connectivity problems, catches remote exceptions, and hides networking from the application;

- can implement Listeners and/or broadcasting;

- throws exceptions for implausibility, out of sync, etc.;

- removes RO clients not responding for a defined time span;

- can stand by, save status to disc, replay, shut down, etc.

The ROApp

- warns and disqualifies misbehaving ROs;

- throws game exceptions, like an illegal move exception;

- supplies a `Game` object for initialization;

- provides references to `ServerObjects`, if the `RealObject` complies with the scenario.

12.3.2 Communication Sequence

Figure 12.2 shows the collaboration of the four main ROApp components. The time flows from top to bottom—from launching the ROApp to shut-

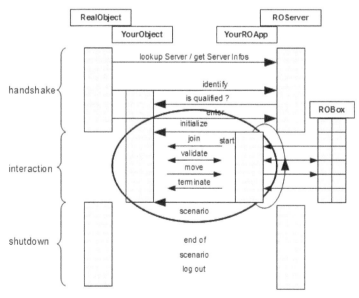

Figure 12.2. The four-layer architecture (see Figure 11.1) in a generalized sequence diagram.

down. The circle indicates the semantic scenario implemented inside the pre-implemented ROServer.

12.4 Application Layer: `RealObjectApplication` Extends `RemoteObjectsServer`

The first step toward writing your own `RealObjectApplication` is simply to extend the server to your own application class. The generalized construction looks something like this:

```
public class RealObjectApplication
                   // or specific app name like LondonChase
     extends RealObjectsServer
// implements RemoteObjectsServer // implicitly
     implements RemoteObjectsApplication
                // or specific app name like RemoteLondonChase
     { ... }
```

We are implementing a simplified version of the game *Scotland Yard*, called `LondonChase` and it *is a* ROApp. Thanks to object-orientation or reusability, even this empty class can already be instantiated inside the `ServerEngine`.

The application designer can override (and hide) the original methods defined in the `roaf` classes, for example, like `LondonChase.identify` overrides `RealObjectsServer.identify`.

Client call: `identify( RemoteObject client )`. The pre-implemented `ROServer.identify` (page 147) method basically receives a remote reference to *any* real object and performs some checks (synchronicity, bounding box, previously logged in, etc.), creates a server-side object, and **puts** it in the `ROBox`. This general server behavior is useful, and it wouldn't make sense to rewrite it in the application. `LondonChase.identify` makes use of this general method via the **super** class

```
// first let ROServer do its part:
   ServerInfo response = super.identify( client );
   if ( response.info == ServerInfo.ENTRYAPPROVED ) {
     // implement application logic and
     // modify response accordingly
     return response;  //   modified ROApp response
   }
   return response;      // unmodified ROSrv response
}
```

This way, the server makes sure that the client is some kind of `RealObject`. If the server denies the entry, the application should also reject it by simply propagating the server's response. If the server approves the RO client, the application can look it up in the `ROBox`. The problem is that

`ROServer.identify` creates a `ServerObject` for the client, but `LondonChase`
`.identify` has no reference to it. This problem was avoided on page 149 by
mapping the client directly to its SO. Therefore, the ID is retrieved in two
steps:

```
// 1. look up ServerObject via client
   ServerObject so = connectedClients.get( client );
   ...
// 2. get identity again
   Object ID = so.getIdentity();
```

The server object should be equal to a server object stored in `candidates`,
which can be verified via:

```
// 3. look up ServerObject via ID
   ServerPlayer
        player = (ServerPlayer) roBox.getCandidate( ID );
// 4. assertion
   if ( !so.equals( player )) { ... }
```

So, in the end, the server implementation is a great help for the application
designer. But how did the `ServerPlayer` get in there?

12.4.1 `ServerPlayer` Extends `ServerObject`

If you recall the pairing of the four-layer architecture,

```
       RealObject    |          RealObjectsServer
         /\     ---+---              /\
   SpecificRealObject  |   SpecificRealObjectsApplication
```

you can tell that the application knows more about specific real objects
than the server could ever know. Looking at the four-layer architecture, one
might conclude that the server can not deal with specific objects created
especially for an application. It would be helpful (and clean) to create
a server-side player to hold specific information in addition to the real-
object attributes; this can be achieved with OO! Just like the identification
method, the creation of a server-side object can be overridden.   This time
the clue is *not* to invoke the super method:

```
// RealObjectsServer implementation:
   protected ServerObject
       createServerObject( RemoteObject client )
   { ... return new ServerObject( client, server ) ... }

// RealObjectsApplication implementation:
   @Override // .. and hide super.createServerObject
   protected ServerPlayer // is a ServerObject
       createServerObject( RemoteObject client )
   { ... return new ServerPlayer( client, server ) ... }
```

This way, the *server implementation* is hidden and unreachable for external clients by the *application implementation*. Although an overriding method has to use the same method signature, the returned object is only required to be *a* `ServerObject`, like any extension of it.

This *covariant return type* is a trick which makes use of software reusability. Note that even the pre-implemented identification method of the ROServer invokes the overriding method of the application without a code change

The `ServerPlayer` is constructed in `RealObjectsServer.identify` by invoking `LondonChase.createServerObject`.

At this point the client player is identified, has a corresponding server player, and the application can finally hand over a `GameSet`. This "bag of things" can include things (objects), like tickets and barriers, to be used in the game scenario. For a minimum implementation of LC, the players should stick to a given network, the game board, or programmatically a `GameMap` (see Section 9.4). The actual network is stored in the file `LDN.net.osm` on the server side. The server creates *one* `GameMap` for *all* players and is free to validate the players' motion against this map. Note that the server consumes time to create this PSF just like the data transfer, which can become problematic for full coverage maps.

The game set is created individually for every client and has to be `final` and `Serializable` to be transferred:

```
// prepare gameSet: load GameMap, role specific stuff,
//                  initial station, etc.
   final GameSet gameSet = new GameSet();
   gameSet.gameMap = navMap;
```

After this identification process, the client can be validated with game parameters, like a valid game role, and is then equipped with a map. The client is free to simply set a valid node as its current position to enter the game. Or, if it is already moving in another scenario (or stand-alone on the local computer), it has to navigate and steer to a valid node to allow continuous motion throughout the virtual world.

Server callback: `init( GameSet gameSet )`. The implementation of the `LC.identify` can be concluded. The client has been identified, has a server record, and is ready to receive a `GameSet` from the server. Note that `player.init( gameSet )` is using the remote concept. This time a return value, time, or exceptions are not of interest. So the method is simply submitted with `remoteCall.submit()`. In order to let the identification finish before initializing the client, the remote method is deferred by a second.

```
   public ServerInfo identify( RemoteObject client )
   {
      ...
```

```
          player.init( gameSet );  // defer init call ...
     ...                           // continue identify
     return feedback;              //   finish identify
  }                                // ... initialize client
```

12.5   ROF: The RealObject Framework

By following the client call `identify` and server callback `player` `.init(gameSet)`, the architect has gained a feeling for the client-server dialog. The `ServerObject` was introduced to simplify the actual application scenario. With the execution framework, the SO is able to buffer client calls and provide the result without blocking the application flow.

`ServerObject extends RealObject` indicates that it also *is a* real object as is the remote object. Both are actors of a real-object application. Technically, they are counter parts to each other describing the *what* and *how* of the client implementation.

A sketch reveals the `ROF`, the `RealObject` framework that is built around the `RealObject`

```
                server                        client
                ------                        ------
            RealObject     <  compare  >
     OBSERVER    /\                        \    CONTROLLER
            ServerObject  < remote API > RealObject
------------------- /\ ----------------------- /\ ---
application: ServerAppObject < remote API >  AppObject

      WHAT? - ServerPlayer        HOW? - ClientPlayer
```

While the client RO implements actual behavior, the SO is dedicated to validating the clients' actions. From a programming side, this constellation can be seen as a generic *controller observer* pattern. The client RO has full control over its behavior, while the SO can use its local RO to reflect, compare, reproduce, and communicate to the server logic.

12.6   ROApp clients: LCPlayer Extends RealObject

The server construction of a ROApp scenario begins with

```
LondonChase extends RealObjectsServer
```

and the associated ROApp clients[3] are constructed with

```
LondonChasePlayer extends RealObject
```

---

[3] In the long run, application and client development should be decoupled to gradually become more and more independent.

No matter how complex and sophisticated the technical ROAF components
are, the ROAF architect should always try to make design choices that
simplify RoApp layers and allow for rapid development of the scenario
semantics. Thus, the `LondonChasePlayer` serves as the abstract base class
for the implementation of a concrete player to actually play the game. The
base class hides client-server connectivity, initialization, etc. The extending
player should focus on playing a smart game and not be distracted with
technology.

Every player class can (has to) implement its own strategy to win the
game. During construction time, the developer should always be able to
run the player client for testing and to troubleshoot problems. And every
player should have a `main` method with the arguments to its home directory
and properties file to begin with.

```
abstract class    LCPlayer extends RealObject
   public class RandomPlayer extends LCPlayer
```

The `LCPlayer` is an abstract template and requires a concrete extension
to supply its name, role, and identity to actually become a player. The
quickest and minimum implementation is a `RandomPlayer` to simplify the
actual (intelligent) decision process to choose the next move. Developers
can use this player as a starting point for their own player and gradually
make it smarter.

```
public    LCPlayer( gpsInfo, name, role, ID )
public RandomPlayer( gpsInfo, name, role, ID )
```

The cascaded constructors set the necessary attributes to participate
in the game: `gpsInfo` satisfies the `RealObject`. Other attributes are stored
in the `LCPlayer`. Valid roles are predefined in the `RemoteObjectsApplication`
interface. Since the game only requires two roles, the integer values are
hard-coded, along with the human-readable strings:

```
public static final String[] ROROLES = {"Chaser","Runner"};
public static final int CHASER = 0, RUNNER = 1;
```

The method `setIdentity( Object ID )` is overridden with a `final`
method and does not allow a change of ID after it is set via constructor (a
way to hide an inherited method):

```
private class RemotePlayerClient
       extends RemoteClient
   implements RemotePlayer
```

The inner class `RemoteClient` was introduced in Chapter 10 (page 133)
to implement methods of the `RemoteObject` interface. And we just (see
page 174) used the covariant return type to override the returned class with

a subclass. The same mechanisms can be used to extend the `RemoteObject`
interface to a `RemotePlayer` interface. The interface adds game-player to
the server-object communication. The `RemotePlayer` implementing class
`RemotePlayerClient` is private and can therefore not be extended. The
connectivity information is hidden to the actual players.

To activate the players remote interface, the method can be overridden
in two ways:

```
protected RemoteObject getRemoteObject()
protected RemotePlayer getRemoteObject()
```

Whatever signature you decide on, the method should definitely be implemented to create the `RemotePlayerClient`; without it external clients will
only get a `RemoteClient` API. When using the first signature, the returned
object has to be cast to a player; alternatively, the casting should happen
inside the second method's signature. And, finally, to prevent renaming of
the super method to `getRemotePlayer()`, the label `@Override` will force the
compiler to fail, which is helpful for the developer to maintain an overview.

Adding client-server methods. Now, the client interface is in place and
the developer can add methods as needed in `RemotePlayer` and then in
`LCPlayer`. The general technique will be demonstrated with the first method
to `initialize` the client player by passing a `GameSet` over the net.

First, the method signature has to be defined in the remote *interface*:

```
public interface  RemotePlayer extends RemoteObject {
    void init( GameSet gameSet ) throws RemoteException;
}
```

Then, the method is implemented in the inner *class*:

```
private class RemotePlayerClient
        extends RemoteClient implements RemotePlayer
{
    @Override // throws RemoteException
    public void init( GameSet gameSet )
    {
        LCPlayer.this.init( gameSet );
    }
}
```

to invoke the `init` method in the actual `LCPlayer`. Note that the methods
`RemotePlayer.init` and `LCPlayer.init` share the same signature for the developer's convenience; nevertheless, they are different methods in different
classes to the compiler.

Next, `player.init( gameSet )` is invoked from `LondonChase.identify` and
as listed in the server callback on page 174. Beware that this `player` refers
to the `ServerPlayer` object, which has to propagate the information to the
client.

Note that the actual implementation `LCPlayer.init( GameSet )` is accessible for extending players. It can be overridden (don't forget `super.init`) to get a reference to the gameset for fancy computations. This is especially important for graphical front ends to get precise geographical coordinates for drawing. A simple player does not need to deal with it and can leave it covered (not overridden).

The next remote method after `init`

```
public int getRole() { return LCPlayer.this.role; }
```

is completely hidden to the extending player, since it does not access a method, only a member of the outer class. The role has to be chosen at construction time for the total runtime just like the identity in `ServerObject.getIdentity()` (which could be subject to change with a deeper identity check). This solution reduces the network traffic.

At this point, the basic mechanisms and initial methods for the client server dialog are implemented and the game developer can add more game methods.

Now it is time to connect the player to the game server. For an exclusive LC implementation, the server connection could be established with the player construction. In the ROAF context, the RO clients can exist on their own to enter and leave ROApps as they choose. Therefore, the client can be hooked to the server in three steps, which enables the implementing player to explicitly decide *when* to take those steps. Note that the remote application is private and inaccessible for the player implementation.

The client first has to look up the server application with

```
final protected static boolean
        lookupROApp( String host, int port, String name )
{
   ...
   rmiURL = "rmi://" + host + ":" + port + "/" + name;
   roApp = (RemoteObjectsApplication)
           RMI.lookupRMIURL( rmiURL );
   ...
}
```

and then the player can invoke `identify` on the server:

```
final protected boolean identifyPlayer()
{
   ... roApp.identify( this.client );
}
```

The server responds by calling back the client with `.init( gameSet )`. Note that the keyword `this` always refers to the full implementation of an instantiated object. It does not and can not refer to the base class only. Therefore, the method does not need to specify the actual implementation of the `LCPlayer`.

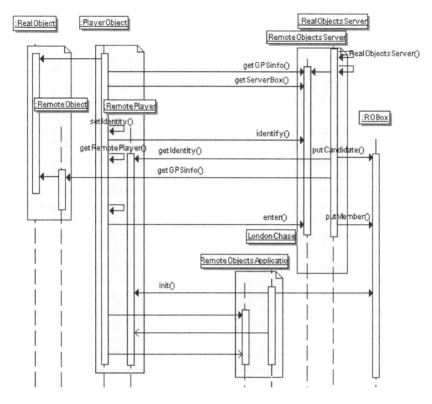

Figure 12.3. The four-layer architecture with a sequence of methods.

After the initialization, the player is ready to enter the scenario, the game:

```
final protected boolean enterScenario()
// =  protected boolean enterGame()
{  ... roApp.enter( this.getRemotePlayer() ); }
```

Before the architect turns to the actual game implementation (wait for all players to enter and start the game) and implements the player moves, he creates an overview of the first phase in a sequence diagram (see Figure 12.3).

The diagram shows the four-layer architecture in a sequential manner with time flowing from top to bottom. The four layers are `RealObject` and `LCPlayer` on the client and `ROServer` and `LondonChase` on the server side. Each class has a corresponding remote implementation for call backs. The construction reveals that the initial communication, like identification with `.identify` on page 147, is pre-implemented between the `RealObjectsServer` and `RemoteObjectsServer` and, respectively, `RealObjects` and `RemoteObjects`. The new extending objects on the client and server side kind of squeeze into this existing dialog. The `LCPlayer` can catch server invocations to override them and propagate them to the `RealObject` (or not).

The diagram describes client-server dialog in its time sequence. In addition, the server can propagate (more or less) useful information during a scenario (game). The Bingo application used UDP broadcasting to inform clients about server events and the concept was analyzed on page 161. For the LC game, the client invocation `.init( gameSet )` is mandatory to play the game. Other information *can be* of interest to the clients and should therefor be broadcast in addition to the vital game information:

> `notifyActors( ServerInfo ) > newServerStatus( ServerInfo )`

Since broadcasting can be useful for any ROApp, it makes sense to integrate it into the ROServer and RO client. The server always has access to all actors of a scenario (which is actually its main purpose) and the `ROClient` is always related to a server. The command `LondonChase.enter` notifies all actors that have previously entered about a new actor with `notifyActors( ServerInfo )` to invoke `newServerStatus( ServerInfo )` on the `ROClient`s.

And, again, the overriding method in `LCPlayer` can receive a special `ServerInfo` for a specialized scenario, in this case a `roa.ldn.all.GameInfo` known to server and clients. Finally, the method `LCPlayer.serverStatus` `Update` dispatches the information to local (abstract!) methods.

This concept is useful for development, since the designer can simply define a game event like `NEWPLAYER STARTGAME NEWMOVE` and wrap additional information in the message string instead of introducing new methods on the remote interface. For example, knowing that the `NEWMOVE` message always looks like

> `"round/player/station=1/0/56"`

the client can easily parse the numerical values and dispatch information to local methods.

12.6.1   Mirror Objects and Object Reflections

Another interesting method (it is dispatched from `serverStatus` `Update`) is `lookupRemotePlayers()`, invoking

> `String[] playerURLs = roApp.getPlayerURLs();`

to get a reference to other players of the game with

```
RemotePlayerInfo player
  = (RemotePlayerInfo) RMI.lookupRMIURL
    ( "rmi://<server>:<port>/<ROApp>/<player><ID>" );
```

Now what are `RemotePlayerInfo`s? Technically speaking, a real object is running on its private host, can identify, enter, and exit one server, enter another one, etc. In addition, the object *can choose to* publish itself on its host:

```
rmi://ROHost:port/myRealObject
```

This is not used in the game scenario for a simple reason: imagine one of the chasers having a direct reference and access to a runner's API. The chaser would skip the third party to validate and propagate game info, and the runner could not be hidden by the application—which is vital to play the game according to its rules. Nevertheless, you might create a *team of players*, who have direct access to each other to coordinate their moves and motion. The server wouldn't know.

Recall that the `LCPlayer`, being a `RealObject`, implements the `Player Client` to realize the `RemotePlayer` API as required to play on the game server. On the server side, the `ServerPlayer`, also being a `RealObject` communicates with the `RemotePlayer` during the game, while its `RemoteClient` is not being used yet.

The server (developer) can use the `ServerObject.RemoteClient` to publish a remote object reflecting the client's behavior. Any client can look up the remote object by its URL and frequently sample it with `getGPSinfo()`. This is exactly what each `LCPlayer` does to get information about the other players. The server can now focus on controlling the game, while each client can actively retrieve the current status as needed.

On the other hand, the server can not steer this additional object, since the server object can only propagate its client's actions; also the server does not have access to the client implementation. In the context of `LondonChase`, the `PlayerInfoClient` is used to provide relevant game information via the `RemotePlayerInfo` API; for example, the game role of each player.

Generally, a ROServer with the address

```
rmi://ROServerHost:<port>/ROAppName
```

can also choose to publish its `ServerObject`s, for example,

```
rmi://ROServerHost:<port>/ROAppName/ROName123
```

to indicate where the ROs come from, belong to, and are validated by.

The server accounts for his published clients. In `LondonChase`, all server (!) players of a game are published and are looked up by its `LCPlayer`s— including their own reflection! This way the player can easily compare the behavior propagated by the server to its actual behavior, and the player becomes the controller of its server player!

If the player gets the current `GPSinfo` via its reflection, the server should immediately[4] invoke `getGPSinfo()` via the `ROClient` (pass through). The

---

[4]Almost immediately, depending on the network latency.

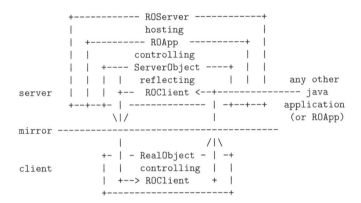

```
            +------------ ROServer ------------+
            |            hosting               |
            |  +---------- ROApp ----------+   |
            |  |         controlling       |   |
            |  |  +---- ServerObject ----+ |   |
            |  |  |  |   reflecting       | |   |   any other
  server    |  |  |  +-- ROClient <--+------------- java
            +--+--+- | ------------- | -+--+--+   application
                 \|/                |           (or ROApp)
  mirror -------------------------------------------
                 |                  /|\
            +- | - RealObject - | -+
  client    |  |   controlling  | |
            |  +--> ROClient    + |
            +--------------------+
```

player can decide to exit the application, if the reflected player deflects from its own behavior. The client marks the server to be untrustworthy.

From the game scenario, you can conclude that the ROClients of the RO and SO refer to the same real object. On the other hand, they (can) fulfill different functionalities.

Figuratively speaking, the SO observes the RO via the ROClient, while the RO can observe and compare its reflection via the SOs ROClient. The ServerObject is a mirror to its RealObject.

12.7 More ServerObjects

Another way to look at the ROF, the RealObject framework is depicted in the diagram on page 182.

Added value by third party. In the larger context of a ROAF, the ROApps act as RO controllers in all kinds of scenarios and add value to the ROs published as output of the server. A server can be seen as a *certification authority* for any RO as it guarantees a well-defined behavior. Vice versa, the number of acting ROs can be seen as a recommendation to others. In a game, the server is responsible to provide equal chances to all players— all players *trust* the game controller. With ROApp *certificates* for ROs, the ROAF can implement a *public key infrastructure* (PKI) to become a network of real-world plausibility.

A RoadServer might implement validations of client traces against a high quality up-to-date digital map and publish them live for applications of any type. A TrafficApplication could use the published traces of the RealCars to show a live map on the internet, received in a car by a navigation system. While the ServerObject is watching the RO client, it itself is being watched by the ROServer against a rule set. Vice versa, if a `RealObject` claims to be real and live, any real-world server should be satisfied with its behavior.

TODO: Server strategy to publish SOs without URLs. For a quick realization of the game, the method `LondonChase.enter` creates and publishes the `RemotePlayerInfo`:

```
...
RemotePlayerInfo playerInfo = player.getRemotePlayerInfo();
playerURL = settings.getServerName()
            + "/Player" + player.getIdentity();
if ( RMI.registerRemoteObject( playerURL, playerInfo ))
{
// player.setRegisteredURL( playerURL ); //other concept ...
   playerInfoAddresses.add( playerURL );
}
...
```

This implementation is sufficient for clients to look up the players on the server's registry:

```
private void lookupRemotePlayers() {
    ...
    String[] playerURLs = roApp.getPlayerURLs();
    ...
```

The developer should understand that after looking up the player info, the client has a direct reference to the remote server player. This reference *is not* disconnected, if the server decides to unbind the object from the registry! Also, the server can pass a reference without publishing an URL.

Another aspect to be aware of is the fact that a `RealObject` does not necessarily know its own URL. Therefore, the server *can* use `RealObject.set`
`RegisteredURL` to associate an object to a URL and use `getRegisteredURL` to retrieve it. The RO might not even be able to identify itself with the URL, and long term, it might make sense to build a URL validation into the RO. For the time being, the strategy is implemented in LC, while long term the ROServer should provide a (filtered, approved, ...) list of published objects. The server can pass references in two ways:

```
        String[] RMIURLs    = RemoteObjectServer.getActors();
   RemoteObject[] remObjects = RemoteObjectServer.getActors();
```

Once a client has a direct reference, the server can only disconnect it by unexporting the `ROClient`.

Acting Server Objects

Up to this point, the `ServerObject` was introduced as the controller to the remote object, with the option to forward the controlled data via its remote client. The client object is acting, while the server object is controlling and reacting, i.e., the `RealCar` client is driving, while the `ServerCar` can assist and propagate warnings, which is also known as `ADAS`.[5]

The Java Tutorial trail RMI points out another option. A remote application can actually transfer an entire object to the server and run it on the server—without revealing the implementation.

An RO developer could program a (smart) player for a (larger) game scenario and transfer the entire player to the server. This way the server processor could replace the client's runtime environment, the developer could shut down his computer overnight and see how his player did on the next day. Then, he could get the player back on his local machine, improve it, recompile it, etc.

This additional implementation makes a lot of sense in the context of a real-world simulator, since it does not compete with current very fast computer games. Think of a game on the German Autobahn where cars try to catch each other, as in the LC game. The distance from north to south can amount to 1,000 kilometers, which would take a car five to ten hours to traverse depending on its type and the traffic (broadcasted via TMC). In this case, it would make perfect sense to transfer `YourCar` to drive for some hours, then transfer it back to your computer and take over the steering again.

For `LondonChase`, the intelligent players are yet to be developed. As an example, look at `LCPlayer.init` and think of a server-side player. The initialization would *not* serialize the `GameSet`. It would pass the reference to the server's game set and use it like any other local object.

Local Server Agents

The Autobahn scenario can be used to introduce another type of server object. At the opening of a scenario, you can't expect hundreds of clients to enter at the same time. For the first clients, the scenario can be pretty boring without other clients to interact with. You can't chase runners before they actually enter the scenario, while a runner would always like to observe a scenario before entering. The solution are actors running on the server side.

If the scenario has a small number of real cars on the Autobahn, the server can trigger a *car factory* to fill the gaps according to the car's speed. These additional cars are smart enough to keep the proper distance and adjust speed to traffic. Population control would leverage the ratio of client and server actors.

`ServerObjects` introduce even more variations to create real objects. An external ROServer should not be able to distinguish a `RealObject` from a

---

[5] Advanced Driver Assistance System.

`ServerObject` by its behavior. Technically, the RO might run stand-alone on a PC, while the SO is running in a server environment.

Consequently, the server can sneak in local server objects created locally inside the server logic. How about a `PoliceCar` pulling over speeding cars? In the case of a game scenario, the local objects can serve as agents to observe the client objects and report any misbehavior to the server logic.

Update: distributed real-world application. Prototyping is learning. In the process, the architect can and should revisit the initial vision once in a while. From the introduction, analysis, and implementation of `Server Objects`, a ROAF law can be derived, which was loosely introduced earlier: *Each RO can only connect to* one *server at a time*. This law is a technical directive and aligns with the design of the Java language, where each object can only inherit from one other object.

Semantically, this law is important to avoid conflicts in the simulated world. Due to the design structure, where ROApps are combined into *one* ROAF, any actor should only exist once. Otherwise, one RO could enter two different ROApps (i.e., in London) and their outputs, the SOs could be combined in a third ROApp. Then, the initial (driving) RO would face its clones making identical moves. The server would have to disqualify at least one of them, the warning would go all the way to the RO, which would have no chance to react or to survive. Nevertheless, this law does not inhibit multi-server scenarios since every RO can have multiple (cascaded) SOs.

12.8   The ROApp Scenario

12.8.1   The `RandomPlayer` Client

The identification and authorization handshakes are now in place. Since the abstract `LCPlayer` can not be instantiated, the `RandomPlayer` serves as an erratic client with a minimum (dumb) implementation to participate in a game. The `main` method serves as the development environment to create players, identify them, enter the server scenario, and play. The developer can tune probabilities for every method, introduce delays, and raise complexity with every development step. A `RandomPlayerFactory` can continue to produce new random players with random IDs to randomly identify and re-enter a game over the complete server uptime. Since `RandomPlayer.main` takes care of registering the client, the random player instance can catch the server callback by overriding the initialization to trigger the game entry:

```
public void init( GameSet gameSet ) {
    super.init( gameSet ); // initialize LCPlayer ...
    // ... trigger player initialization from server call back
    // set random station and enter scenario ...
    // delay method return during development
```

```
    enterScenario();
    ...
}
```

Due to the decoupling of the client server, the developer can start the server and then start random players on one or more JVMs.

We now go back to the details of the game at the state where the first player-client has registered and entered the game.

12.8.2 Game Controller

The player factory in `RandomPlayer` creates a number of players to enter the game. On the server side in `LondonChase.enter( client )`, you can find a `Controller` instance to represent the game master. The controller should always have control over all game state and the involved classes. Technically speaking, the controller serves as a monitor and is created with the properties file.

The controller should be the main authority to dispatch the properties to the application's components. In other words, the controller synchronizes the game activities, and the vital methods have to be `synchronized` to ensure clear states of the game. The developer should be aware that synchronized methods are much more costly and should only be used, where necessary. Generally, the `synchronized` keyword does not belong to the method signature and can also be used on a higher overriding layer.

Once all players for one game are identified and have entered the game at a station, the controller sets the list of authorized players, creates a `Callable` to notify all players about the game start, and starts the game—after returning from `enter`:

```
if ( allAboard ) { // let's play
   controller.setAuthorizedPlayers( players );
   Callable<Void> startGame = new Callable<Void>() {
      ...
      notifyActors( status );
      ...
      controller.startGame();
      ...
   };
   executorService.submit( startGame );
   // submit and immediately continue
}
```

The controller changes its status from `REGISTERING` to `PLAYING` and notifies all classes waiting for something to happen, for example, the `GameThread`.

A `GameThread` to run rounds and take turns. The `GameThread` is `started` with the server construction and waits for the controller to start the game. Once the game starts, it retrieves several settings relevant to each game and then starts the game cycle for each round and all players in each round. The running players get the chance to move (to a safe place) before the chasers start moving.

In every round, every player is required to announce his destination (a valid station on the map) where he is going to head next:

```
destination = currentPlayer.getDestination( time2move );
```

This `ServerPlayer` method is not just getting the identity of a player. In this case, the player *must* announce his next move to remain in play. Therefore, the client call is blocked for a well-defined time to move (in the properties). If the player fails to submit the next destination, the game branches off to the `catch` block of the `TimeoutException`. For a quick implementation, the game designer can end the game in this case.

12.8.3   ROAF Reconciliation

The rule "every player is required to announce his destination" may sound a little fuzzy in the context of a computer program. The reason behind this rule is the *reconciliation* of the board-game constraints with a potential real-world scenario.

When playing the board game, each person (i.e., operator) moves his physical player from one station to a neighboring station during his turn. For the board game, it makes no difference, whether the player is actually moved or being picked up and set down on the destination. The only thing that matters is the valid destination. On the other hand, the ROApp needs to be designed without conflicts with real-world actions. Therefore, LCPlayers can not simply jump from origin to destination. They really have to move there and the task is to reconcile motion and duration within the game rules.

The game rule dictates that the chasing players have to "hit" the running player at a valid station. Since a computer needs precise instructions, this means that the four `double` values describing the `Position`s of a runner and a chaser have to match precisely. This is hard to achieve, since `double` values are internally coded with binary fractions and can not always represent decimal fractions precisely. So even parsing a decimal number from a string can vary the actual value. This problem can be overcome by introducing a tolerance of a few meters. This tolerance agrees with the fact that you have to walk a few steps from a subway station to catch a cab. Even better, each station should be defined by the area (polygon) defined by the subway's platform, the bus station, and the taxi stand. One question remains: How fast should the players move? In the real world, the question is "How fast *can* a cab move conforming to traffic rules, traffic lights, and the traffic itself?

Reconciliation analysis. Here are the basic steps:

1. In Round 1, the first runner *starts* moving toward his destination. The game master can validate *how* he is getting there by validating the motion against the edges of the navigable map at a reasonable frequency.

2. After the first player starts to move, the next player (chaser) has secondsPerMove to pick a strategy and start moving.[6] Then, all the other players start moving one after the other. When the last player has *started* moving the next round begins.

3. *Before* the first running player begins to move again, all other players potentially still moving need to get a chance to capture him.

Consequently the server has to check, if a chaser is headed to the runner's destination and delay the invocation of the next move.

Note that the LCPlayers can navigate internally, i.e., adapt speed, while the navigated objects only replay a given trace.

Move implementation. Every LCPlayer has two integers myDestination myOrigin to define his current move. The player is asked to move by the GameThread with the method getDestination() and must submit the next destination after a maximum of getTimeToMove() seconds.

```
PlayerClient.getDestination() > LCPlayer.getDestination()
```

The PlayerClient invokes the *abstract* method getDestination() on the player. Each player implementation can choose a way to return the destination (randomly, analytical, intelligent, interactive etc.).

```
public Integer getDestination() throws RemoteException
{
// start timer
   submitDestination = LCPlayer.this.getDestination();
   move(); // start move and (re)set origin and destination
   return  submitDestination;
}
```

Then, the player actually starts moving by invoking the RO method move(). Due to the restricted number of valid moves, the LCPlayer can pre-implement discrete moves from a valid station to a neighboring station and make it final.

---

[6]The name of the *board game parameter* is a bit confusing and could be renamed to secondsToDecide.

```
final protected void move()
{
    Route route = board.getLink( myOrigin, myDestination );
    int secsToMove = settings.getTimeToMove();    // [s]
    double distance = route.getTotalDistance();   // [km]
    double speed = distance / secsToMove * 1000;  // [m/s]
    moveOnRoute( route, speed ); // returns immediately
    myOrigin = myDestination; // for next route calculations
}
```

The move should be validated before submission to retrieve the `Link`, which is a `Route` from the digital map. With the given route, the player can calculate the overall speed according to the game logic.

In the process of the game design, a new method is introduced to the `RealObject`. The method `moveOnRoute( speed )` is an interpolation between `move( dir, speed )` and similar to the playback thread. The idea is to tie the RO's motion to a fixed route with implicit triggers to change direction and simply accelerate or decelerate the object according to environmental constraints. While the playback is completely automated and placed in the `GPSunit`, this method is part of the RO.

Note: A `Route` does not describe an actual path. It is only a sequential collection of geographic points. As long as the route is driven as the crow flies, the total distance can be calculated. As soon as a route is fed into a navigation system, the system calculates a path along roads of the map. Different systems can use different algorithms and map data resulting in different paths, duration, and distance.

Chapter 13

Mission Accomplished:
Time to Play

13.1   Introduction

At the end of the last chapter, the *architect* completed the book's mission
defined in Section 3.8: `LondonChase` is a client-server application with the
clients representing players populating a server application (game) with a
map scenario (game board) and a rule set (controller).

This entire chapter contains hands-on instruction for developers and
development groups. It marks the end of the implementation phase and the
beginning of the exploration phase. The reference application implicitly
prototypes the `roaf v1.0` library and marks the end of the initial coding
from scratch and the beginning of an evolutionary life-cycle development.
Developers can work on different details of *one* application with the common
constraint to keep it up and running.

Besides providing a "minimum implementation with maximum abstraction," the application development has finally reached an entry level for
end *users*. The game is in a state where non-programmers can participate
in the evolutionary process, by playing and providing feedback.

13.2   Using the Application

For the reader of this book, it has been a long way to a distributed application. Yet, why all this fuss about programming techniques for a simple
board game?

Learning Java with the *London Chase* client-server framework can be
fun—even for beginners. The book can be used to introduce most aspects
of programming in Java and object orientation. For classroom use, students
can be split into teams to drill down into different components described
in dedicated chapters and in the Java Tutorial Trails.

As stated earlier, the application only meets the minimum requirement
to run out of the box. This partial functionality of a lean application is

much more tempting for beginners to find room for improvements. To provide some ideas, this chapter will include a number of problem statements for useful additions and features. You should also check the listings in this chapter with the actual current state of the software at `www.roaf.de`, home of the ROAF developers.

13.2.1   Setting Up the Environment

The entire distributed application resides in the folders

```
... \workspace\roaf\ ...
... \workspace\resources\london.roa\client
... \workspace\resources\london.roa\server
```

The software was developed in the Windows path `D:\virtex`, which is hard coded in different files and needs to be adapted to the environment.

Running the server. Please replace the absolute paths in this batch file to reflect your environment:

```
... \london.roa\server\LondonChase.bat
```

Next, open the server policy file and also adapt the path:

```
... \london.roa\server\ROServer.policy
```

For a quick start of server and clients on one PC, the RMI policy is set to `permission java.security.AllPermission`. As soon as clients and server run on different hardware, the policies should be restricted as described in the RMI trail. After the two files have been modified, the server software should be launched from a command line shell with `LondonChase.bat`. If everything works out, the screen prints the different steps of the server startup process:

```
open prop file: ... \ROApp.props
-- listing properties --
    :
Created registry on //localhost:1099/
    :
allocated registry on //kbeigl-acer:1099/
remoteObject bound to ROserver at 20:55:03 CET 2010
RemoteServer is registered: //kbeigl-acer:1099/ROserver
load osm network file: ... LDN.net.osm
    :
mapGraph is setup for navigation.
GameThread was started ...
    :
Start waiting for players to register for game ...
```

At this point, the server is up, running, and waiting for clients to register.

Running random client players. The client batch file is easier to adapt, since the policies are not used to run on the same computer as the server. The batch file

```
... \london.roa\client\RandomPlayer.bat
```

is a regular Java invocation with two parameters (separated by a space!). Please replace the absolute paths in this batch file to reflect your environment, open a second console, and run the batch. By placing the two windows next to each other, the client-server dialog can be observed:

```
server> notifyActors: Game will start in 15 seconds ...
client> STARTGAME Game will start in 15 seconds ...
server> ********** round: 1 **********
client> round/player/station=1/0/134
```

The *maximum* duration of the scenario is roughly defined in the `ROApp.props` file:

```
secondsPerMove x nrOfRounds = 60 x 22 = 22 minutes
```

Note that the server is *not* terminating after `Terminating Game` ... and has to be killed manually.

The main method of the `RandomPlayer` client makes use of a *player factory* for development purposes. Many clients can be created with this one method to test the server's behavior for various combinations of clients.

Of course, the server and client can also be launched from an IDE. In order to set the RMI policy and command-line parameters, they have to be added to the "Run configurations ..." (i.e., in Eclipse) in the fields "Program arguments" and "VM arguments."

Now everything is set to run the application on two JVMs and to modify the code for a clear view of the scenario.

13.2.2 Distributed Programming and Deployment

The concept is simple. First the server software is set up. Then each programmer or programmer team is asked to install the client software (and IDE) and run his personal version of the `RandomPlayer` as described in Section 13.4.1. Now, the environment is setup and the games can begin.

Every student can start making his player smarter and join a new game any time. In this chapter, we collect a small initial TODO list and share it online at `www.roaf.de`. The problem statements are enumerated for reference purposes and most of the implementations can be found in Parts II–IV.

Problem 13.1 (Software distribution.) Separate the packages into four Java archives: server, common, client, and GUI. Install and run server and common jars on the server (without client jar). Install and run client and common jars on the clients (without server jar).

In a classroom situation, the class can be split into server and client developers and receive server/common and client/common packages, respectively. Since server and client software are separated, they can be developed independently. Each team can appoint a GUI developer to create a GUI for server and client software, respectively.

Naturally all GUI developers can again form a team. In the best case classroom scenario, there is a classroom with a connected PC pool and a beamer. Since the beamer can serve well as a visual front end for the whole class, the GUI team can attack the following problems.

Problem 13.2 (Create a server GUI.) The `main` method of `roa.ldn.server` `.ServerEngine` is prepared to create a GUI with `srvEngine.createGUI()`. Add a server window showing the server's bounding box, the network, and the moving players.

A solution can be found in the `Navigator` software. Of course, there is much more a GUI developer can come up with. Problem 13.2 is important for the collaboration of all teams. Ideally, there is a beamer connected to the server machine and everyone in the room can watch the current server and game status. Note that the server GUI should *not* display a map image, since the game is actually restricted to the network.

The `RandomPlayer` represents the *minimum implementation* of a player. A programmer can take the player as is, set up the game server, and run a first game round to make sure it works. Then he can start modifying the player, run another game round, implement decision making, run another round, etc.

`RandomPlayer`s can always be used to fill in missing players until new players start playing against each other. Finally, the developed player can go online to challenge other players. With this development cycle, the player can be validated at any time and provides motivation to drive the development.

The ROApp developer can also configure `RandomPlayer.main` for any number of required players and roles to improve the server logic step by step. Clients and server can run on different JVMs connected by the network. For the client player, the server should be set up to run on a separate computer to start, end, and restart games over the entire development session. The developer should not be able to and should not have to touch the server at any time after its startup. Instead of modifying the server code, the developer should create a "must have" list and communicate it to the server team; that is the main idea and advantage of distributed application development.

In a game of only `RandomPlayer`s, the developer has a hard time to really know what's going on. He could direct the players' positions to the `System.out` on a console, but that is not really helpful. Also, it would be a waste of time to wait until two random players actually meet at a game station. Therefore, the first player implementation should be a `GUIPlayer` to serve two purposes. The game should be visualized for intuitive observations, and it can be used to invite real people to interact in the game and challenge the automated players.

Problem 13.3 (Create a client GUI player to interact with the game server.) Extend the `LCPlayer` to a `GUIPlayer` and add a GUI front end showing the game board. The user should be able to participate in the game via mouse and keyboard.

As an interactive GUI player is really useful, we implement Problem 13.3 in Section 13.3 and also provide some guidelines for Problem 13.2.

Note that classroom sessions should be restricted to the `ldn` packages, while the `roaf` packages should be recognized as a given library (and can also be isolated in jar files). Before modifying the library, the developer should read the final part of the book for an overall understanding of the ROAF directions. At this point, we should only be modifying the game, not modifying the ROAF.

The server team can start out with the file `LondonChase.java`, which is reasonably short and introduces the application logic. This team is responsible to keep the game server up as long as possible and announce down times in advance. A first task could be the following:

Problem 13.4 (Create a server cycle.) Modify the application logic to wait for players to start a game. When the game is over, or interrupted, the server should allow `candidate` players to enter a new game.

Here is another task that would be helpful.

Problem 13.5 (Modify game parameters during server uptime.) For effective development cycles, the server administrator should be able to delay a game start, modify game parameters, and then start the game.

Besides improving the game software, the server team can also act as a game referee for unexpected situations. Competing players could be analyzed by examining their source codes or even added to the server logic with a local server player. There are many such tasks and the server team will certainly devise some of their own. Nevertheless, here is another mandatory item:

Problem 13.6 (Implement game over.) If the chaser gets caught the game should be finalized.

In addition to the server and GUI team, two main teams should compete in runner and chaser development. Section 13.4 will demonstrate the basic operations for intelligent decision making—again with minimum code to get developers started. The chaser developers should be aware that *Scotland Yard* is a cooperative game, since a single chaser can never catch the runner alone. The developer can create teams of players communicating during the game to share algorithms and coordinate a strategy.

The UDP broadcasting technique is described in Section 11.9. After each coding session, the players should compete in a few games to gain some experience and observe opponents. Finally, teams can compete in a well-defined tournament. Note that these tournaments can also be held via the internet with the developers sitting at home.

This chapter only suggests a concept for team development on a predefined game application. Nevertheless, it is vital for the ROAF development to distinguish between the game and the ROAF vision. Before investing too much energy in the game, one should read the final part of the book to make sure the useful improvements for the ROAF will become part of the official online version.

13.3   Interactive `GUIPlayers`

An interactive GUI player enhances the development process since it can be directed by human intelligence. The player allows its operator to actually *see* the game and to let him *decide* where to move. The player extends (assists) the operators eyes, hands, and brain.

This section details the client-server process at the application level and can serve as an approach to other player implementations. First of all, the `GUIPlayer` *is a London Chase* player:

```
class GUIPlayer extends LCPlayer
```

This construction combines the GUI and the player in one class. Two separate classes would require a third class to propagate information between them. The main method reads the game settings from a local file, creates player and GUI, and launches the player on a the local JVM. The developer can choose to supply an initial property file or to store the file after the first game. In addition, the player is supplied with a game map to support real-world orientation.

Make sure that the server (i.e., `ServerEngine`) is up and running to accept new players. Once the `GUIPlayer` is running, the GUI provides the necessary triggers to sign up at the server and play. For this special game, the bounding box is predefined and the corresponding map image is loaded. It can be displayed by clicking the checkbox.

When the user has clicked the button "connect server," the server is looked up with the `settings` via the pre-implemented `LCPlayer` method:

```
lookupROApp ( host, port, name )
```

The object variable roApp is set internally and remains hidden to the player, yet the player can access predefined server values, for example,

```
mapArea = getServerBox();
// -> LCPlayer.roApp.getServerBox()
```

which is basically the bounding box of the given game board for this very special implementation.

Now, the GUIs box can be readjusted to the server box:

```
gui.  bigMap.setGridArea( mapArea );
gui.smallMap.setGridArea( mapArea );
```

The button "register player" invokes

```
identifyPlayer()
```

to submit the internal (and also hidden) remote player with the desired game role. Then, the server calls the client player back on its init method

```
public void init( GameSet gameSet )
{
   super.init( gameSet );
   createLinkArrays    ( board );
   createStationArrays ( board );
       ...
```

where the LCPlayer (super) is initialized before the overriding GUIPlayer method retrieves the board to create all links and stations of the map.

Due to the rather preliminary design of the MapPanel, the player is forced to manage a number of drawing arrays. Two arrays to draw the map links

```
private Route[] mapLinks;
private Color[] linkColors;
```

by predefined feature colors

```
private Color[] featureColors = // ( R, G, B )
{ new Color( 204,153,28), new Color( 34,141,95),
  new Color( 209,13,85) };
```

Another simplification to the ROAF is a single map (or network) for server, all players, and for the duration of a complete game. The arrays can be set up with fixed lengths. An application without a defined bounding box, like cars driving on interurban roads, will require a different and dynamic approach to constantly read and buffer the map for routing and drawing.

Two checkboxes are supplied to show or hide map links and game stations.

Before the player can enter the game scenario, the rules require him to set his origin as the entry point. The actual variable myOrigin is defined and used in the abstract `LCPlayer`—but is not being set there. To make the user's life as easy as possible, the `GUIPlayer` implements a map click to find the closest station:

```
stationID   = board.findClosestStation( click );
stationPos  = board.getNodeMap().   get( stationID );
stationName = board.getStationMap().get( stationID );
```

and a double click sets the player's origin and the underlying real object's position:

```
myOrigin = stationID;
setPosition( stationPos );
```

At this point, the player can enter with the "enter game" button to invoke

```
enterScenario();
    ...
    roApp.enter( this.getRemotePlayer() );
```

and the server immediately echoes

```
serverStatusUpdate > "NEWPLAYER ROserver/Player10"
```

After entering, the player has to wait for other players to enter until one game is loaded. For a straightforward development, this status is volatile and the user should simply wait for the notification via

```
serverStatusUpdate > "STARTGAME" > gameStart()
```

This can then be achieved by launching another JVM with the `RandomPlayer` `.main` method or with additional `GUIPlayers`. Although each new player is announced, the quick implementation waits for the game start, before creating fixed-length player arrays with `createPlayerArrays()`

```
private      int[] plyRadius;
private    Color[] plyColors;
private Position[] plyPositions;
```

with the predefined colors

```
private Color myColor = Color.BLUE, runColor = Color.RED;
private Color[] playerColors = {
    Color.DARK_GRAY, Color.BLACK, Color.CYAN,
    Color.MAGENTA, Color.PINK, Color.ORANGE, Color.GREEN };
```

Since arrays have a given order of ordinal numbers, an index is added to look up the array index by player ID:

```
/** Map player ID to index of fillPositions. */
private Map<Object, Integer> player2index;
```

Note that the array length adds one entry for the local player, which should match the remote player reflected by the server. Then, the player positions can be set with

```
updatePlayerPositions();
```

Before the server starts to announce the player's moves one by one, the destination arrays can be created and positioned outside of the map area with

```
createDestinationArrays();
    private      int[] destRadius;
    private    Color[] destColors;
    private Position[] destPositions;
```

The destinations that should be displayed with **drawPositions** raise a conflict with the station arrays created earlier. Both arrays have constant lengths, are merged with

```
mergePDArrays();
    private      int[] allRadius;
    private    Color[] allColors;
    private Position[] allPositions;
```

and are managed with

```
checkedShowStations();
```

At the end of **gameStart()**, a **Sampler** thread is created to update all player positions asynchronously to the actual game logic.

To assist the user in decision making, a single click on a station displays all possible paths from the origin. By clicking on intermediate stations, the paths can be reduced until a neighboring destination is displayed on the map. A double click on the destination submits the move to the server and starts the move.

Just like the destinations and stations have to be merged to be displayed, the arrays

```
    private Route[] routes;
    private Color[] routeColors;
```

compete with the digital maps arrays and have to be managed by a GUI logic.

Now, one or more users can play a game by simply clicking on the map.

Here is another mode that would be helpful:

Problem 13.7 (Implement an observer button.) Add a button to observe the game instead of participating in the game.

13.4   Intelligent Players

Artificial intelligence (AI) is *the* challenge for programmers; usually the problem is to find a feasible environment to experiment with. Although the ROAF was defined *not* to implement any intelligence, it does provide a perfect framework to experience the effects of decision making. In the long run, any `RealObject` has to have some intelligence to find its way through the real world (applications).

Implementing a `RealTrain` in the ROAF is not a great challenge. The train could easily be simulated by supplying its actual schedule and the digital map data of the corresponding tracks. A `RealBus` is a little less predictable, since it does drive according to a schedule, but also has to deal with the traffic on its route. A `RealCar` (or cab) has even more freedom (unpredictability) than the train or bus. It is restricted to streets (solid, even surface); within these limitations, it has a *choice*.

Artificial intelligence can *only* make choices within the given degrees of freedom. In *London Chase*, each player has to move on the digital map's links and only has the choice between the direct neighbors of its current station. The chaser should choose the best path to the runner, while analyzing the runner's behavior and predicting his next move. The runner should basically do a similar analysis, albeit with opposing conclusions.

Depending on the size of the network and the number of players on each side, the strategy can become complex and time consuming to implement. Before implementing the actual decision making, the game can be defined in terms of heuristics: a graph, a tree, search algorithms, etc. Then, each individual player can decide *how* to search. AI literature provides hundreds of approaches including Nilsson, A-star, minimax algorithms, alpha beta pruning, by breadth, by depth with (variable) weights and functions for nodes and edges, etc. The reader should choose his favorite algorithm!

The game of *London Chase* is the ideal (beginner's) environment for intelligent and dynamic route calculations. The AI player introduced in the following sections shows one way to interpret the game. In the long run, the environment will get more complex and the algorithms need to be adapted accordingly to give the RO a good orientation in any situation it may face.

Note that a navigation system is using complex calculations for the current position and route. Nevertheless, most people would not follow the calculated route in their own city. People sense even more parameters than the system.

13.4.1   Creating a New Player

Here's a quick start guide for creating a new player that we will designate as `AIRunner`:

1. Open the file `RandomPlayer.java` with a text editor (not in the IDE yet).

2. Copy the content into a new file (does not need to be saved).

3. Replace all `RandomPlayer` Strings with `AIRunner`.

4. Open the IDE and create a new class `roa.ldn.client.players` `.AIRunner`

5. Copy the modified file content into the new class and save.

6. Save the file `../resources/london.roa/client/RandomPlayer.props` as `AIRunner.props` and add it to the command line to launch `AIRunner`.

7. Modify the main method to `playerFactory(1,0)` as an example.

8. Run a game with the new player (and other random players) for testing.

9. Replace the player factory with a single-player construction in the main method like

```
public static void main(String[] args)
{
    connect2server(args);
    GPSinfo serverMetrics = LCPlayer.getServerGPSinfo();
    AIRunner runner = new AIRunner( serverMetrics, "R100",
                RemoteObjectsApplication.RUNNER, 100 );
    runner.identifyPlayer();
    try { Thread.sleep( 2000 ); }
    catch (InterruptedException e) {}
    runner.enterScenario();
}
```

Now a game can be started:

1. Start `ServerEngine`.

2. Start the new player, i.e., `AIRunner`.

3. Start one or more `GUIPlayer`s.

4. Start one or more `RandomPlayer`s to complete the actors.

Be sure to provide individual IDs, when using more than one GUI player! The GUI players can be used to direct the game.

13.4.2 AIRunner

The player running away from the chasers is a good starting point, since
he always knows the current game status, i.e., the other players' positions,
while the chasers can only see the runner every other round and have to
make assumptions about his status.

With the new `AIRunner` created in the last section, the programmer has
a working client program with only two methods

```
public void init( GameSet gameSet )
protected Integer getDestination()
```

to initialize the player with game information and to actually make the
next move. The rest of the interaction is implemented in the `LCPlayer`.

For a quick implementation, a heuristic evaluation will be placed in
the method `getDestination()` with the time constraint of `getTimeToMove()`
seconds in the server settings. If all players were to use this mechanism, the
game rounds would implicitly indicate the individual decision times. The
runner always begins the evaluation with the chasers' constellation when
he is requested to make his move. The initial station is randomized and
can be hard-coded to analyze special situations.

The decision-making in this simple game context can be based on the
chasers' stations and the board. The runner does not need to distinguish
the chasers actual ID. The `LCPlayer` keeps track of the game and provides
the member variables `destinations players board`. With

```
moveStations = new ArrayList<Integer>
                        ( board.getDestinations(myOrigin,1));
chasersStations = new ArrayList<Integer>
                        ( destinations.values() );
```

the AI runner has gathered all necessary information: the chasers' stations as input parameters, the list of all possible destinations (result set)
to evaluate, and the board itself to validate the strategic quality of each
station.

A heuristic search should store intermediate results in case time runs
out. The initial value is randomized:

```
moveStations.get( generate.nextInt( moveStations.size() ));
```

Since the destinations provided by the `LCPlayer` include the runner itself, it
has to be removed before evaluating. According to the rules, a runner can
not move onto a chaser's station, so those stations can be removed from the
move destinations. Also, the chasers' neighboring stations can be removed.
Otherwise, the runner will be defeated in the next round:

```
removeChaseStations ( moveStations, chasersStations );
if ( moveStations.isEmpty() )
```

```
        System.out.println("I can't move!");
   removeChaseNeighbors( moveStations, chasersStations );
   if ( moveStations.isEmpty() )
       System.out.println("I give up!");
```

If no moves are left, the default move can be executed; if one move is left, it is executed immediately; if more moves remain in the list, the actual evaluation can begin. This minimum implementation sorts out the stations with the most connections and randomly picks one of them.

The runner has learned some rules that can be used for future rounds.

Heuristics This implementation is intended as an appetizer for more games. The player becomes much more effective by continuously evaluating the board and players. Once the **AIRunner** has been launched and registered (not entered yet), it can already start the *first step* by analyzing the game's network provided by the ROApp. Every station can be scaled with the average distance to all others and stored in a look-up structure. Note that the maximum distance on the game board is only nine. This is fine for learning.

In the *second step*, the runner can observe three chasers positioning themselves on the board and then evaluate a good position to enter the game. In a classroom situation, this might represent the first competition, since another runner could enter the game and exclude other runner candidates.

The main task to make the player intelligent is to create one heuristic formula to consider all parameters of a game situation and calculate a weight. In the case of *London Chase* the parameters include

- my own and the other players' current positions;

- the maximum distance to the runner's position;

- the best strategic positions from which to run away;

and a *Scotland Yard* implementation would add

- a determination of network features and the remaining tickets of each player.

The results should be scaled from 0 to 1 (100%) to reflect the quality of a move compared to other moves.

Decision tree. The heuristic function represents the players' intelligence, and it can be fine tuned after every game round, or it can be programmed to learn during the game. In either case, the function can be used to create a decision tree. Once the game is in progress, the runner can internally start a separate thread to continuously extend and maintain the tree. The root node represents the runner's current position and its child nodes reflect the next possible moves. Then, every move can be scaled with the heuristic and the best move is chosen.

A further development cycle could introduce game triggers to the tree thread, which drop the unapplied branches, move the executed move to the root, and supply the latest game parameters. In the end, the runner is continuously building a tree, and the server request to make a move triggers the selection of the best move.

13.4.3 AIChasers

The chaser template is already prepared as described in Section 13.4.1.

The implementation challenge is different and more complex than that of the runner. For one thing, all chasers form a team and the order of their moves actually doesn't affect the game. First, the runner moves and then *all* chasers make their move.

In order to implement a team strategy, *one* heuristic function can be created to calculate *all* chasers' moves. The function is more complex, since the runner's location is only published by the server every ((round - 2) % x == 0) with x being 3, 4, or 5 for different skill levels. While the runner is hidden, the chasers can create an estimation tree for all possible runner moves.

While the runner tree can search the tree by depth, the chasers' tree is more effective by searching the full width of each round. Before entering the game, the chasers can determine the best positions on the game map to reach any other station quickly and subtract two moves to get there as soon as the runner shows up. One of the main goals is to minimize the total distance, which is the sum of all chaser' distances to the runner. Additional criteria can minimize the escape options and avoid different chasers from blocking each other. Yet another aspect to consider is the (hidden) time consumption of route calculations. The results of each calculation could be stored in a matrix to access previous results and avoid redundant calculations.

Finally, a technical solution has to be found to communicate each calculated move to the relevant player. For example the search thread could be placed in one leading player, who broadcasts the move to its followers.

With the normal *London Chase* parameters of one runner, three–four chasers, and 199 stations, the calculation time of one minute can be a challenge. The tournament rounds can begin with two minutes and slowly reduce the time to move, which will segregate the better players. The game complexity can be raised by, for example, introducing barriers to block the runners best escape routes, features for ferries, and tickets for features can be added.

13.5 *Robocode*

Before concluding the initial RO and ROApp development, the reader is advised to download and have a look at the excellent game of *Robocode*, which is the brainchild of Mathew Nelson, a software engineer at IBM. The whole idea of developing, deploying, and playing can be experienced with *Robocode* as a great environment for Java programmers to experience life-cycle development and to anticipate ideas for the ROAF.

One complete life cycle of *Robocode* can be divided into these phases:

1. Download, install, and launch the *Robocode* system.

2. Pick some robots, start a battle and see what it's all about.

3. Open the Robot editor (and a robot template), and code and compile a robot

4. Test the robot against others or against its own clones.

5. Improve the robot and test again.

6. Deploy the robot in jar file for battles.

The `Robot` in *Robocode* is the equivalent to the `RealObject` of the ROAF. It is controlled via its API—the remote control to a robot enabling the programmer to move the `Robot`. The location is described with Cartesian coordinates on a battlefield similar to a `RealObject` located by GPS coordinates. A `Robot` burns energy just as the `RealCar` needs fuel. Collisions and dynamic behavior are pre-implemented (in the base object) and a `Robot` can communicate with teammates.

Part V

ROAF: Real Object Application Framework

The mission of this book has been accomplished, but the vision goes way beyond the specific application.

The current status of the `roaf` framework (library) and the source code and resources for the *London Chase* game are available on the ROAF developer website at `www.roaf.de` as a basic architecture. The software predefines a structure for more and more complex components to support the functionality of a real-world simulation.

In order to streamline further developments, this final part of the book will revisit the vision and provide semantic guidelines for developers to ensure a constructive collaboration even with highly deviating ROApps for many different scenarios.

The mission of the book was to get an understanding of the new system and supply a running sample application. At the end of this development process, the architect has gathered enough knowledge to provide a white paper for the actual roll out of the product.

One of the main purposes of the real-object application framework is to harmonize, synchronize, and overlay different ROApps on a higher level. At the same time, the ROs should become more independent of a ROApp and be able to wander from one ROApp to another—inside the ROAF.

Chapter 14

Evolution

14.1 Introduction

The software developed in this book is only an indication of the ROAF capability; it establishes the first cycle for *evolutionary* ROAF development. It ties together different components, which can be developed almost independently while keeping the application running.

The end of the book marks the beginning of the ROAF evolution. This chapter provides general guidelines for your involvement at www.roaf.de,

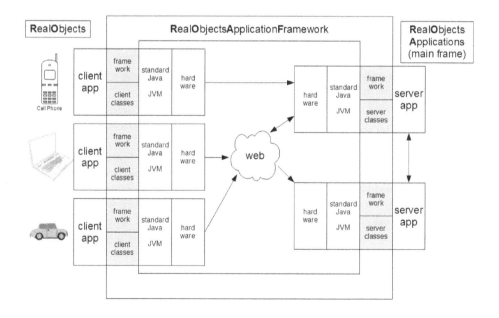

Figure 14.1. The Real Object Application Framework is a standard software library for real-world scenarios with real objects as their actors.

the home of ROAF developers. Many ROApps define *one* ROAF. Before
the ROAF can grow organically in life cycles the reader needs to contribute
his ROApp. The real-world aspect should connect all ROApps together in
the long run!

14.2   Game Scenario versus Real World

For some developers, the *London Chase* implementation might be disap-
pointing and not a great challenge. One runner against three or four chasers
is the player limit for a reasonable game scenario. On the other hand, it
is a feasible size for developers working alone or in a private training environment.

The main reason to choose the game scenario of *Scotland Yard* is the
abstraction or simplification derived from the real world. The game of
Scotland Yard was not implemented to its full extent in order to leave
some room for individual development. For a Java course, the game is
sufficient to study the architecture, add game logic, and develop intelligent
players on the application layer, i.e., `ldn` packages.

In order to pursue the vision, the `LondonChase` application should be
copied as `CityChase` (CC) to branch from the game scenario to the real
world, from London to any other city. Just like the moving and navigating
objects were left behind, the game is only the entry scenario to the ROAF—
the technical proof of the concept.

There are many ways to break out of the game limitations into a more realistic scenario. Depending on the developer's intentions or domain knowledge, there are many aspects that can be improved.

Game Board versus City Map

A most apparent problem of our implementation of *Scotland Yard* is the
tiny *population* of the scenario restricted to four or five players.
The straightforward way to extend the number of players is to create
a London city map with more stations. The map images of London in
`..resources/london.roa/client/pix` from the osmosis zoom level Z15 indicate the network density of the real city.

For a long-term approach, a dedicated map compiler can be created
to extract the original game board network from the real London Open-
StreetMap. This initially involves hand selection to identify the relevant
nodes and edges. Consecutive extractions should recognize the previous
network, i.e., by adding ROAF attributes to *the* OpenStreetMap (see Figure 14.2). It is also a good idea to conserve OSM compatibility to be able
to apply OSM tools and viewers.

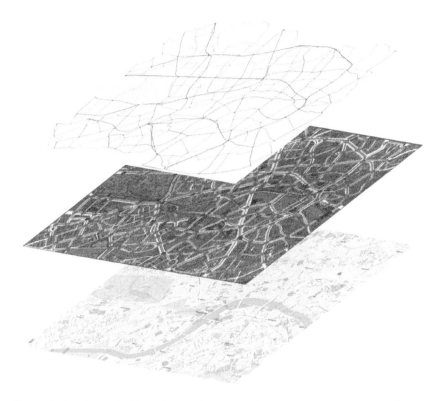

Figure 14.2. Every ROApp has to be reconciled with the real world. A game board should be extracted from a real map with a network for moving objects.

Then, the next cycle can refine the network. A formula (or ratio) to estimate the *population density* would be helpful: x stations are feasible for y players. In the context of a distributed development, the number of stations should be increased gradually, maybe from 199 to 300 in the first step. The development in small steps is important, since the game logic and players' intelligence most likely has to be adapted at a certain, yet-to-be-determined threshold. With gradual growth, each developer can work on one component at a time.

Another aspect of creating a new map network is the original design of the *Scotland Yard* game board. In the game context, every vehicle type has it own network, and the rules don't imply any shared links. In the real world, busses and cars predominantly share the same streets, which requires multiple features for a number of links, and a slight modification of the rules.

Don't forget to share your map and compiler at `www.roaf.de`!

From City to City to Country

Improving the map without changing the software seams to be a good start. From a purely technical aspect, the `CityChase` (CC) application runs with a network that has three features (vehicle types) and about 200 stations. Therefore, another variation from LC is to move the scenario from London to another city; if subway links are to remain part of the game, the city should have a subway.

For the `Navigator` project, Germany was stripped down to its major cities according to their population, and it will make sense in a little while to move the CC scenario back to Germany. The `GER` map also represents a network with three features and 38 cities ($\approx$ stations, nodes). Therefore, the CC software should also work with this map. The country map adds a higher magnitude or different granularity to the game parameters in terms of traveling distances.

From Scenario to Scenario

Of the 38 destinations of the `GER` map, the largest and internationally recognized cities are Berlin, Hamburg, Munich, and Frankfurt. One CC ROApp server could be set up for each city to allow the developer community to test their players in different environments. Each player, running on a private smart phone, PC, or CPU JVM could set the RO properties to a dedicated city, register, enter, and play.

The longer traveling distances on the `GER` map modify the character of the game. The operators (i.e., the people behind the players) can not be expected to watch their RO traveling along the Autobahn for hours without any chance to leave it. The necessity to automate players becomes obvious, and a new type of ROApp can support this.

This additional ROApp can also run on its own machine, with the same software. The technical challenge now is the implementation of a *single sign on* mechanism and generalized entry and exit rules. All CC ROApp servers (`GER`, `B`, `H`, `M`, and `F`) run with identical CC software and have contracts with each other to form one "world." Their networks have to have connectors for players to enter a scenario.

In this server cloud (see Figure 14.3), the player should learn to find his way from city to city. For example the CC rules could require every player to enter at a city *not* having its own CC server. Then, the player can travel to Munich, for example, and the `GER` application would automatically register the player on the CC-M ROApp, since the authentication and authorization has already been done. Once the CC-M hands over the city's network, the player could play games until he decides to leave (or run away from) the scenario and travel to the CC-F scenario. The ROApp servers could set up a scoring system to keep track of every player

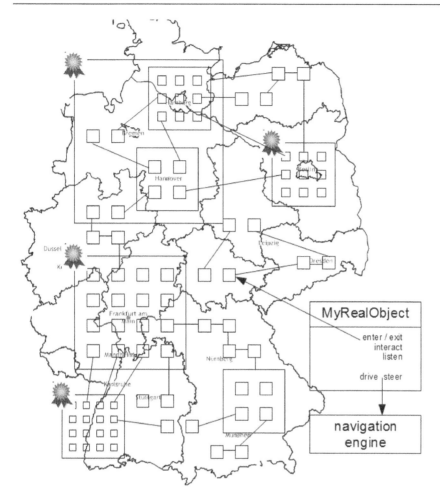

Figure 14.3. A `RealObject` can enter a cloud of collaborating ROApps. The small boxes represent the ROs enclosed by ROApp boxes. Some of the ROApps are certified to be "ROAF harmonized."

(`role/wins/nrOfGames`) and pass the record to the `GER` application for an overview released on a website.

From Discrete to Continuous Motion

In Section 12.8.3, we replaced the discrete moves on a game board with continuous motion. The navigating objects of the `Navigator` move continuously and can be sampled by an observer. Nevertheless, the objects move at speeds of about 10 km per *second* !

Generally, ROAF harmonization requires scenarios to be *real-world compatible*. Only *harmonized ROApps* can become part of *the* ROAF.

In the CC cloud, this has consequences for the player. A player leaving the CC-M server needs to travel hundreds of kilometers to the next CC ROApp. Depending on the transportation, e.g., railway, waterway, or Autobahn, this can take hours. Therefore, the CC rules could reward players entering once and competing in different cities. Vice versa, the players could hardly be operated by a real person all the time.

In the long run, all CC players traveling at realistic speeds become part of the one ROAF and can be displayed on one map together with the HD and RGB scenarios (Chapters 7 and 8), which play back real traces. Again, the rules should be adapted to allow continuous motion instead of discrete rounds with discrete stations. With realistic traveling speeds, the task is basically the same. The dynamic routing would have to be improved to consider traveling times that come close to the runner or runners.

Fractal Design

The idea of harmonizing ROApps is similar to a meshing of map layers. As described in Chapter 7, a map can be composed of many layers showing different aspects of the real world. On the harmonized Germany map, many ROs can move along the network without showing what they are really up to—like watching cars in the real world.

In the real world, public transportation is perfectly scheduled. Similar to the moving objects of the `Navigator`, `RealBusses` and `RealSubways` could be supplied with routes representing their lines and stops and extended to traces with the time stamps of their schedules. To distribute responsibilities, a separate ROApp server could be set up with the schedules of (London's) subways. The server is *closed* for external ROs to register and enter. Instead, the server has an internal RO factory to instantiate `RealSubways` as server objects (SOs) on demand. These SOs can be registered on a LC game server as external ROs, like the players. And, again, the game rules have to be adjusted, since the players will actually have to wait for a subway to arrive at a station to board. The external subways become SOs of the game scenario and can be enriched with their physical dimension, a number of cars with a limited capacity for people.

With this strategy, the servers providing live data for public transportation become an important, reliable, and stable source for other ROApps. While subways don't have to worry about traffic, a server for bus schedules might raise conflicts in the scenarios. On the other hand, many modern public transportation organizations already use live GPS tracking for their fleet control. If these signals would be fed into dedicated servers, every RO would have to give right of way. Otherwise, the server could *tag the RO as*

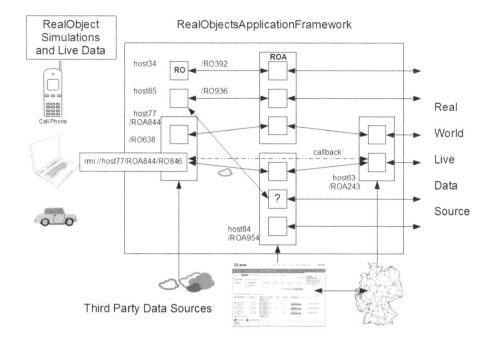

Figure 14.4. Third-party data can be fed into the ROAF via any ROApp. For example, `host84/ROA954` can provide a public train schedule—and the `RealTrain`s.

unrealistic. Public transportation can be seen as a four-dimensional grid of the real world and ROApps with official information can be certified to be real.

With this approach, the CC applications would become smaller and more restricted to their rules. The LC game could be composed of a subway, a bus, and a game server. Another aspect is appealing: the ROs could access third-party sources to inquire about the official schedules, automatically via web services.

14.3   ROApp Classification

Running LC on a PC is the entry to your own ROApp. The development steps in the last section should help you to break out of the game scenario into the real world. At this point, you are the ROApp master defining the rules in your world. You can place information for your ROApp at `www.roaf.de` to invite people to populate the scenario on your host machine. Anyone can set up and host a ROApp to provide ROAF web services. Since every developer has a different domain knowledge, this section provides a number of scenarios as appetizers and some guidelines to keep in mind for the ROAF.

Looking at the OSM map features might inspire your imagination to add your own rules. For example, the traffic lights can be extracted from OSM and come alive in the scenario. A traffic management system could be implemented as a new component of a City ROServer. Then, every participating RO would be required to *listen* to the signals and stop at red. The number and meaning of map features can be implemented gradually.

Autobahn (MRE) Scenario

The `Navigators GER` map is highly simplified and should be refined to the actual road geometry. In Germany, the Autobahn is the highest-level road (Interstate) and represents the network for overland traveling. Like Germany, every country has major roads, and it would not pose any new issues to extract the *major roads of Europe* (MRE) for the entire continent.

Driving across countries takes time and burns fuel. Therefore, a MRE game scenario could introduce a fuel capacity for each `RealVehicle` and implicitly define a traveling range. Why not add gas stations to the scenario? Then the route planing would have to consider gas stations (and opening hours) to travel long distances.

If the ROs want to leave the continent, they will need boats or airplanes. Similar to the MRE network, boots and planes also travel on well-defined routes and would not require any new ROAF technology.

Cab Caller App

The *Scotland Yard* game board has a taxi feature for selected roads. Nevertheless, the game does not really need cab, bus, and subway vehicles; it therefore has to be harmonized to be eligible to be part of the ROAF.

The `CityChase` scenario should introduce the three vehicle types in addition to the players. Since a cab should be available for every player at every cab station, their density should be higher than that of the players; one cab per player is not sufficient, since a cab can't catch up with the player, if he takes the subway. A `RealCab` fleet could implement its own intelligence to spread over the game scenario and cover all cab stations (taxi stands) reachable by any player.

Actually most cab companies are already prepared for a real-world "cab caller." Most cabs are traced with GPS and their signals are available at the cab company's headquarters. These signals can be fed into a cab caller ROApp to show the `RealCabs` driving through the city. The customer could look at the city map and click on the closest cab to call it. Once the driver is informed, the ROApp can return the estimated time of arrival (ETA).

If the customer has a GPS phone, he could use the waiting time to buy something at the next store. When he comes out of the store the cab is standing right there to pick him up; if not, the customer can press "call driver" or check the map for his current position. This ROApp adds only a small layer to the ROAF packages but a lot of added value. The customer can be identified and simplify the payment via electronic cash with another button "pay."

POI Sites

Every navigation system has *points of interest* (POIs) in its data to enable passengers to travel to well-known places by name. POIs have a wide range of categories and have developed a market of their own. While POIs are only names connected to an address and a location, their owners (restaurants, hotels, shops, etc.) have a vital interest to add up-to-date information as needed.

In the ROAF, a POI can be represented by an independent RO (machine). The POI becomes much more than a point on a map. With CAD standards, any building `Shape` can be modeled by extending the `RealObject`. In the Autobahn scenario, gas stations could be built along the road and provide information about waiting times, amount, and price of gas. The `RealCars` would need to take an exit (change lanes) and drive to the pump.

Generally, POIs could be maintained similar to websites. Providing independent POI machines introduces a paradigm change in the world of navigation and location-based services (LBS). A hotel runs its own computer to add 3D information, such as the location of vacant rooms and current prices. A RO visitor could automatically process this information and compare hotels along the road, book the room, and retrieve navigational instructions to the parking lot.

The RO would become a little ROApp.

Real Estate

With the development of navigation systems, the demand for map data has grown massively. Cars with a large number of sensors and 360 degree photographic- or even laser-capturing capabilities, drive around cities to digitize the world.

Accordingly, existing markets have to adapt to this new technology. In the real-estate business, data is collected for every building: photographs, ground plans, etc. This data could be digitized and loaded into `RealHouses` to be available in the ROAF: "the live Web." These houses could be visited or simply enrich the environment, while other houses might only exist as a building footprint with height information.

Multi-scenario Augmentation

The list of ROApps indicates their potential variety. On the other hand, *the ROAF is intended to be a collection of realistic live scenarios.* ROAF-certified ROApps can be augmented seamlessly like a collection of live information. Each ROApp can retrieve as much third-party information as needed and form a new filter to provide this data to the ROAF.

The *London Chase* game is initially unrealistic. Nevertheless, it is a ROAF candidate, since it *can be* gradually harmonized with reality. The players could be real people with a GPS smart phone, and the ROApp could propagate the runner's location selectively. It is important to understand that the naturally growing ROAF can also be used as a web service for fictional scenarios.

Formula 1 racing is very popular across the whole world. Every car is equipped with many telemetric sensors and transmits live. The racing tracks are publicly available with high precision and could serve as a ROApp map for a racing scenario. In a fictional environment, the real cars could be reflected (live playback) as built-in SO racing cars, while an additional race car could be an external RO. Thus, the RO could actually compete with the real cars. Today's racing simulations could be connected to the RO to add real physical behavior according to the car's setup.

Obviously a City application with real tanks driving downtown through buildings and shooting at each other should, while possible, not be released to the ROAF. The ROAF could, however, provide the city map and physical information on streets and buildings *to* this application.

One of the most significant attributes of the ROAF is the concept of *now*. From the developer's point of view, the "switch" is the ROApps GPS clock. As stated on page 169 : "...a ROApp should *not* be set to real time to indicate that it is *not* trying to reflect the real world." So by setting the time to *now*, as every GPS device can do, a public ROApp is staking a claim and can be tested against the ROAF. This processing of live data has many advantages. The most outstanding advantage is to avoid persistence. It's easy to organize a source for weather data and public transportation schedules—for today. Schedules are already synchronized implicitly and can be validated against each other. Simulating any other day in any city becomes much more complex; the implicit timing is another advantage. Live synchronization can even augment live cams in the ROAF world and Formula 1 races can be transmitted by data only, to be reassembled to a 3D animation by various front ends.

The *Inception* Scenario

To emphasize the idea of using reality as a starting point, science fiction movies can be modeled with ROAF technology. The movie *Inception* written, produced, and directed by *Christopher Nolan* and distributed by Warner Bros. Pictures can be modeled as an ROApp.

Inception is about entering another person's dream world (in this case without a hard-wired connection to the brain as in the movie, *The Matrix*). One person is hosting other people's mental self-projections in his dream world. In ROAF terminology, the dreamer is not only providing his RO, but rather comes with his own ROApp. People entering other people's dreams are intuitively recognized by the projections of people close to the dreamer. These projections are nothing but SO agents in his ROApp dream world, which he can model with his own creativity; like bending physical laws.

Inception adds more fictitious ROApp layers to the scenario. In a simple dream, any actor can wake up from the scenario by dying in it. For a dream in a dream (in a dream), a powerful sedative is needed to stabilize the layers of the shared dream. In the fractal ROAF, a dream in another dream can be modeled with one main ROApp. When entering the deeper layer, another ROApp comes up with another scenario; yet, the second ROApp underlies the higher ROApps rules and laws. For example, loss of gravity in the higher level influences the lower level accordingly. This cascade of ROApps adds technical challenges to the ROApp architect.

In each layer, one person (RO) is creating a dream (ROApp) to travel further into the initial dreamer's subconscious. The exit rules are complex; a dreamer has to be "kicked" physically in order to pop out of the current dream level. This kick has to be set up by a trusted person outside of the dream world by kicking the physical body to fall and lose his sense of gravity. Another appealing aspect is the time factor: every dream level has a subjective time factor of about five to 60 minutes (i.e., 1/12).

So, the ROAF should pose no limits to your imagination.

14.4 RO Classification

Just like a ROApp can reflect almost any (real) world scenario, the RO can represent any (physical) object in a scenario. In the long run, ROApp and RO development should be gradually decoupled. Loose coupling implies that individual objects should reduce unrelated information, while high cohesion is complementary and provides information connecting objects to collaborate.

In the ROAF context, the cohesion (framework layer) is provided by reality and scientific knowledge, while individual developments can be loosely coupled (application layer). Constructing ROs independently for different ROApps will make them robust in the long run. The object can be tested in different environments and under different aspects to reduce them to their core attributes and optimize them for reusability.

The general ROAF development is driven by the real-world aspect and the RO hierarchy should classify real physical objects. Things can be grouped together according to common characteristics and separated from things with differing characteristics. The classification should be based on existing scientific organization. The objects in the hierarchy have to be precise to spare developers from coding general knowledge.

14.5   Decouple Mature ROs

The `roaf.ros` package is the heart of the ROAF and its `RealObject` is the main actor of any real-world application. The `RealObject` is the fundamental object with built-in connectivity and GPS traceability under the hood. For a general understanding, the RO and the ROAF are related in a way that is analogous to the `JComponent` in the `javax.swing` *framework* described in Chapter 6.

In terms of real-world programming, the `roa.ldn.client.LCPlayer` is a dead end. The player is dedicated to the board game and is prepared to implement game intelligence. Analogous to the gradual transition from a board game to a more realistic scenario described in the last chapter, the `RealObject` can be gradually decoupled from its environment.

The players in `LondonChase` actually move by using some kind of vehicle. To reflect the real world, the scenario should be composed of `RealPersons` using `RealVehicles` as the base class of `RealTrains`, `RealBusses`, and `RealCabs`, as a special kind of `RealCars`. In addition, the cities providing the environment for the `CityChase` action have to be modified toward reality.

For the game of *Scotland Yard*, this programming effort is a waste of time, since it doesn't add much value to the game logic and strategy. The idea of creating `RealObjects` is the separation of internal and external behavior and a useful prerequisite for distributed development.

The easiest implementation is the `RealTrain` without the freedom to choose a route; it has to follow the rails. Since the `RealBus` is tied to a route, it has to react to the traffic. The `RealCab` can basically navigate through a city and additionally receive customer locations to pick them up.

Besides creating new `RealObjects` for the `CityChase` scenario, the ROs can also collaborate and interact. For example, a `RealTrain` can be created with a number (instances) of cars. Inside the train, a `Conductor` could walk up and down the train and ask people (players) for their tickets. Again, this would require `RealPersons` to get on and off the train, call a cab, etc.

RealObject.Shape

A GPS signal alone does not provide any information about the object being traced. The idea of the individual `RealObject` programming environment is to add as much information about the object as possible, like the telemetry of a racing car. With CAD technology, every object can be modeled to have dimensions. For cars, this information is important to avoid collisions. For buildings, this modeling process can be even more complex. At a certain population level, it might make sense to add a train station. The station has different entrances, a number of platforms and rails. Since the stations usually belong to public transportation services, they might as well run on their own machine to ensure the logistics to their guests.

A use case could be a `RealPerson` getting out of a `RealCab` at the main entrance of the `RealStation`. The person would walk into the station and immediately get access to a list of all trains arriving in the next 30 minutes. The person would walk to the platform, wait for a `RealTrain` to arrive, get in, buy a ticket, and find a seat.

In the ROAF context, the station is an object (RO) to access public information and to host `RealObject`s of many different kinds, e.g., like a list of all trains passing through, a list of all other people in the station, etc. Any person could ask any other for local knowledge.

14.6 The `RealObject` Hierarchy

Since a `RealObject` can be extended to any thing in the real world, each client should be `classified` (`instanceOf`) or upcast to a well-defined `roaf.ros` object. This provides access to additional specialized methods and allows the server to extend this object to an SO with the same API.

Figure 11.1 shows the `roaf.ros` package, which can be extended to the RO hierarchy (Figure 14.5) on the client side. Nevertheless, the package is *not* called `roaf.clients`. In order to create more specialized real objects, it makes sense to also install RO packages on the server side. Then, the client objects can be *classified* more precisely, by using reflection mechanisms on the server side, for example. Pairing the client and server process becomes part of one `RealObject` development, and the developer should take care to code the framework in a way that methods can be added smoothly in one place, before they spread over the entire application.

Revisiting the `RealObject` Generator

After the exploration process, the idea of a `RealObject` generator described in Section 3.3 should become much clearer. When looking at OO documentation, one could conclude that objects can be developed isolated from

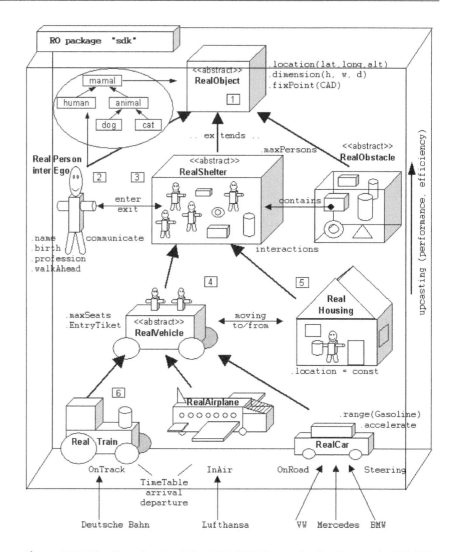

Figure 14.5. The first sketch of the initial RO hierarchy leading to the ROAF.

applications. One of the main OO ideas is to use predefined objects to populate and develop an application.

A `RealObject` generator could be based on the results of a search machine for different combinations of terms. One word combined with different words can represent different perspectives of a term. Then, the search results could be analyzed and combined (subset, superset, intersection, etc.) to provide a list of useful and obligatory internal and external attributes.

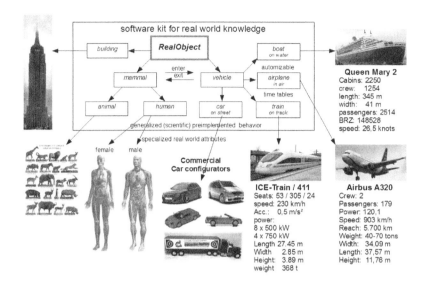

Figure 14.6. Any RO hierarchy can be modeled with a number of `roaf.ros` packages.

The most used objects tested in different scenarios will gradually gain robustness, quality, and become more realistic and (each consecutive version) learn to survive many scenarios and become an certified member of a RO hierarchy (see Figure 14.6).

`RealObject`

The RO is the root of the hierarchy (or hierarchies) and intentionally designed to be **abstract**. The actual values for mass and shape have to be provided with the extension and instantiation to a real thing. Technically, the RO provides the basic mechanisms for a client-server handshake to identify and authorize the client for the server application.

In the long run, it might be necessary to provide mechanisms to provide values interactively, when signing up on a less known server application, for example. Different scenarios can apply different rules (and laws) to address dedicated attributes (and methods) of the actors. Isolated ROApps (i.e., those not certified for the ROAF) could "switch on" gravity to observe the objects falling in the gravity field; the mass attraction on Earth should vary from a moon scenario.

RealPerson

Modeling people is probably the most complex RO of all. The experience is, however, most interesting, when guiding a residual self-image as in the movie *The Matrix* through a virtual world. In the times of social networks, a `RealPerson` generator for a dedicated site (i.e., format) becomes obvious. The basic idea is that people don't want to enter personal information over and over again. They have their personal account to store this information and retrieve it as needed.

A generator could retrieve standard information like name, birth date, and home address from a personal profile to instantiate the initial person. Once instantiated the operator (real person) could be asked to provide other relevant information like email, phone number, etc. The personal identity (digitized with a certificate) should be secured, since no one would like to publish passwords, credit-card data, or a personal medical history.

If we like it or not, people become more and more traceable by digital footprints and a motion profile provided by cell phone and GPS. The biologic identity is supplied by the genetic fingerprint although it is still too time consuming to analyze DNA on the fly. Who knows when a new technology emerges to identify personal sediments instantly to reconstruct a person's path through every day life and look him up in a database.

The physical body. Religions claim that all people are equal before God. In constitutional states, all people are equal before the law. For a scientist, people are only equal in one single aspect: every person has the same normal body temperature no matter, if he is in Alaska or on the equator.

On the other hand, the human body is very well known to physicians. i.e., the step length can be derived from the body dimensions. Therefore, the `RealPerson` template *can* be programmed in a general way, *before* the actual values are supplied.

A good first approach to a validation of a human body is to look up sports statistics. World records mark the explicit maximum performance and should at least raise a warning, when being broken. Science can also provide tables for fuel consumption. How much energy (calories) is burned with different activities? And activities can be combined. For example, it is most unlikely (impossible?) to set a marathon world record within a triathlon competition. Similar to the fuel capacity of a car, a person has to fill up his tank and a ROApp server could observe a person and require him to navigate to a place to drink and eat and to sleep.

The brain. The brain can be another software component of a `RealPerson`. This part of the body can be easily defined, physically. The mental processes are almost impossible to access. Nevertheless, any person can be observed in a social context to reveal his attitudes. This external view (from the ROApp) can be interesting to simulate social behavior. In artificial models, simple theses can be implemented to simulate a population and observe the effects in the community. In the end, we only see other people externally no matter how close they are.

RealVehicle

The programming of a `RealPerson` seams pretty vague at this point and should be kept in mind for future study. There are much more promising physical objects to work with, like vehicles. A vehicle is 100% man-made, and a blue print exists for every one. In OO, the blue print is mentioned often as the initial idea of a class.

While people are gradually beginning to digitize their identity, most vehicles are publicly described in detail. Similar to personal profiles in social networks, most car vendors supply a configuration tool for different models. The consumer can choose a vendor, then the model, fuel type, and so on. At the end of the process, a `RealCar` generator could create the RO composed of a well-defined body, wheels, windows, doors, and engine. This process is comparable to the composition of a Java Bean with dropdown lists of predefined values. Also, images of the car can be added to depict it from different perspectives in a scenario.

In contrast to a real person, most cars are perfectly predefined and the ROs could be created for every release of a new car. As described in the development of GPS traces with a Buell motorcycle, the external observer can validate the actual behavior. With all of the information about cars, the server can classify a car (top-down) step by step: A high acceleration excludes cars with little power; a narrow street between buildings and a low bridge restricts the car to smaller models. The SO reflection of the car can temporarily store these values as they are determined. A configured car can serve as the root class to personalize (i.e., tune) it and as a reference implementation on the ROApp server side.

Traffic and interaction. Once instantiated, the car can be traced and saved to a history. In the ROAF, even the history files could be validated against a road network and traffic rules. A little ROApp server scenario could be used as a driving school and a more complex one could require a certificate by this server as a driver's license.

Driving a car in an isolated environment is good for learning how to drive. Yet the actual challenge is to guide a car through the traffic at rush hour. In Germany, the server can make sure that the car is always driving on the right side of the road. In order to automate the vehicle to drive programmatically drive, the car needs to be equipped with sensors (i.e., Listeners in Java terminology) and has to react to events propagated by the server.

RealObjectContainer

The `RealPerson` and `RealVehicle` were described as well-known examples of the most common objects in our everyday life. Figure 14.6 provides the general idea for constructing well-known commercial objects. Most objects can be constructed from an internal and external view to be interpolated in the ROAF world.

Technically speaking, there should be `RealObjects` to hold other objects. A car can hold a limited number of passengers, a train can hold more passengers distributed over the wagons, etc. Again, the Swing framework described in Chapter 6 can supply the idea. The `java.awt.Container` class is a component containing and partly managing other components. In Swing, these components are arranged by the container class. In the ROAF, a container class can aggregate the populating `RealObjects`. For an elevator, the number of passengers is only one relevant attribute, while the sum of all masses can be vital for a normal operation.

At the moment there is no construct for a `ContainerRealObject`; it is a challenge to the architect to supply a general way to construct one. A `Container` interface might be a good approach. Or a little ROApp as one `RealObject`!

Appendix A

Downloads

The website `www.roaf.de` is the home of the real-object application framework (ROAF). The site will be constantly evolving with your input and will reflect the ROAF development beyond the book's mission of providing a "minimum implementation with maximum abstraction." This implementation should encourage the reader to develop his own scenario and establish a life-cycle development of the `roaf` library in a community process.

You should visit the site on a regular basis to find out what other readers are doing with the software. You are kindly asked to also report any problems you may encounter. All problems, errata, and links will be referenced to the appropriate sections of the book.

Since the book describe the development of a library and a sample application, the source code with the relevant resources should be downloaded and compiled *before* reading the related chapters. The source code was developed with the Java Developer Kit version 6 (`JDK 1.6.0_ ...`).

In order to download the ROAF sources, you will need to sign up with your email. Your email will not be used for purposes not related to this book. Please understand that the sign-up procedure provides important project statistics. With your ROAF user ID, you can also join in discussions on the ROAF Forum and provide helpful feedback on the ROAF Wiki to cooperate in creating a professional `roaf` software. Of course, all contributions will be collected in a contributors' list to make you an official ROAF developer. After your registration, you can download the sources from the website section "sources & resources."

Basically *all* resources of the book are organized in the three top-level folders `book roaf resources` which can be downloaded to one `workspace` root folder:

```
<yourPC>\workspace
      +-- book
         +-- src          ~250 kB
      +-- roaf
         +-- src          ~500 kB
      +-- resources       ~92 MB !
         +--- deployed    ~120 kB
```

```
+--- gps              ~47 MB !
+--- london.roa       ~12 MB
+--- OSM.compiler     ~33 MB !
```

A.1   Source Code for the Exploration Phase

The `book` folder should be your first download: `book.src.zip`

The archive holds the code developed in Chapters 1 and 4–10. The entire folder can be checked into your IDE as one project and the compilation (and Java documentation) with the Standard Java tools should add `bin` and `doc` folders automatically.

```
../book
  +---bin
  +---doc                   first occurrence in
  +---src                   chapter.section
  |  LinkCompiler.java        8.3.2
  |  NMEAconverter.java       4.12
  |  OSMparser.java           7.8
  +---roaf
  |   +---book                4.2
  |      +---gps              4.2
  |      +---intro            1.4.1
  |      +---map
  |      |   +---gui          5.4
  |      |   +---osm          7.10
  |      +---navigation       8.4 + 9.4
  |      +---rmi              10.2.1
  |      +---ro               6.2
  +---roafx
      +---gui\map             5.5
```

In order to test the compilation, you can run the `main(String[] args)` method of the `MassObject` class in the `roaf.book.intro`. This is the first method to be executed in the course of the book in Section 1.4.1. The simulation should print these (and more) lines to the system output (`System.out.println`):

```
moon is at position 382679989
earth is at position 0
  :
simulationtime: 41 seconds
  :
```

Since the reader is expected to be familiar with the basic concepts of the Java Language and Java tools, the book will refer to methods in a brief format: "Please execute the method `roaf.book.intro.MassObject.main`."

The `MassObject` class does not have any dependencies to other classes and should run as is. Nevertheless, the dependencies of other packages and classes developed chapter by chapter are constantly increasing, and, therefore, the book packages are delivered in one download archive.

After the exploration phase, which ends with Chapter 10, you may want to remove the `roaf.book` packages as they are independent of the release `v1.0` of the actual `roaf` library and the `roaf.roa` sample application (see below). Or you can choose to create two projects in your IDE to make sure not to introduce any dependencies between them.

Note that the `roaf.book` packages are deprecated and therefore stable for book studies. So please remember, if you consider contributions to the ROAF, you should first study the `roaf.book` packages and play around with them. Beginning with Chapter 11, the release candidate for `roaf v1.0` will be introduced. This release will be published under version control.

A.2   Resources

All resources *can* be downloaded in one archive: `book.rsrcs.zip`

The resources will be the second download you will need and you should download these files at the latest when you begin Section 4.4. They can be downloaded from the "book resources" webpage. Note that this archive has more than 90 MB. The overall folder structure and sizes are indicated on page 228. The resources are subdivided into four main folders:

```
                              first occurrence in
    ../resources                chapter.section
       +---deployed                  5.8
       +---gps                       4.4
       |    +---GER                  8.4
       |    +---HD                   4.4
       |    +---RGB-BUELL            7.10
       |    +---RGB-BUELL-NW         6.4
       |
       +---london.roa                11.2
       |    +---client               11.5
       |    +---server               11.4
       |
       +---OSM.compiler              7.3
            +---deliveries           7.4
            +---products             7.10
            +---tools                7.4
```

Like the `roaf.book` packages, the resources are also deprecated. Both download archives will only be modified if major problems occur or if their usage can be improved. Remember that the website should support development beyond the book and is open for your contributions. In the course of the book, you can develop your own version of the software and share similar archives to the ones provided.

Hint: Sometimes the source code indicates the path originally used, for example, `nameIn = "D:/workspace/resources/gps/HD/HDcastle.gpx"`. You can modify these paths to your environment or use the general path variables defined in the `roaf.book.Misc` class. `Misc` determines the `APPATH`, when it is loaded into the JVM and sets the individual resource paths.

A.3   Release Candidate `roaf v1.0`

The `roaf` folder is the final download required: `roaf.src.zip`

This archive marks the end of the book's development and the starting point for the `www.roaf.de` community. Actually, the archive consists of the three top-level packages `roaf`, `roafx`, and `roa` representing the `roaf` library for client and/or server, the `roaf` GUI, and the reference implementation of the ROApp `LondonChase`, respectively. For the time being, these main packages will be distributed in one archive; their functionality is rolled out in Chapters 11–13. Nevertheless, the architecture supports independent development and the application layer should be branched off as indicated in Chapter 14.

```
../roaf
  +---bin
  +---doc
  +---src                    first occurrence in
      +---roa                chapter.section
      |   +---ldn
      |       +---all (common)  11.5
      |       +---client        12.6
      |       |   \---players   12.6
      |       +---server        11.5
      +---roaf
      |   +---all     (common)  11.4
      |   +---gps               4.2
      |   +---roa               11.5
      |   +---ros               11.3
      |   +---util              4.2
      +---roafx
          +---gui
              +---map           5.5
              +---panels        11.4
```

A.4   Third-party Sources

Since most of the third-party sources are listed in the text, you should visit the `www.roaf.de` website, if the links or the software have changed. In the section "third party" you can find all links to external sources; these will be updated on a regular basis. Due to licensing and updating policy, you should download the third-party software from these external links. All of

the listed software is free and open source. Please visit the `www.roaf.de`
website to find

- The Java Developer Kit,

- The Eclipse IDE,

- OpenStreetMap tools described in the text,

- JavaGPS,

- *Robocode*,

- ...

The reader should also be aware that the release candidate `roaf v1.0`
marks the end of the book and is subject to (your) changes. The text
already suggests some third-party resources, like libraries for GPS, spatial
indexing, and a graphical front end. These external libraries have not been
decided on yet, and the website may also serve as a place to discuss and
validate concrete implementations. Naturally, all external sources have to
harmonize with the software license of the resulting `roaf` software.

Index

Printed and bound by CPI Group (UK) Ltd, Croydon, CR0 4YY

21/10/2024

01777097-0002